W9-API-302

Nature Incorporated

TS
1323.15
.S74
1994

NATURE INCORPORATED

INDUSTRIALIZATION AND
THE WATERS OF NEW ENGLAND

Theodore Steinberg

University of Massachusetts Press   Amherst

WOODBURY UNIVERSITY LIBRARY

DISCARDED

Copyright © Cambridge University Press 1991
First published 1991
First paperback edition, 1994 by
The University of Massachusetts Press
All rights reserved
Printed in the United States of America
LC 94–15035
ISBN 0–87023–943–0

Library of Congress Cataloging-in-Publication Data

Steinberg, Theodore, 1961–
    Nature incorporated: industrialization and the waters of New
England / Theodore Steinberg. — 1st pbk. ed.
        p.   cm.
    Originally published: Cambridge, England; New York: Cambridge
University Press, 1991.
    Includes bibliographical references and index.
    ISBN 0–87023–943–0 (pbk.: alk. paper)
    1. Textile industry—New England—Water-supply. 2. Textile
industry—New England—History—19th century. 3. Water rights—New
England—History—19th century. 4. River engineering—Environmental
aspects—New England—History—19th century. 5. New England—
Industries—History—19th century.   I. Title.
TS1323.15.S74   1994
333.91'23'0974—dc20   94–15035
                                                                CIP

British Library Cataloguing in Publication data are available.

Publication of this book is supported, in part, by the New Jersey Institute of
Technology.

BOOKS ON DEMAND
This edition has been produced using digital technology that
allows the publisher to reprint small quantities "on demand."
The textual content is identical to that of previous printings,
but reproduction quality may vary from the original edition.
Printed by IBT Global, Inc.

To Helen and Madeline
and in memory of Seymour

CONTENTS

FIGURES, MAPS, AND TABLES

Figures

Maps

ix

Tables

PREFACE

Thus far, mine has been a metropolitan life, one spent in urban and suburban sprawl where the natural world is largely obscured by the asphalt, steel, and concrete that weigh down the American landscape. Nature seems mostly absent from this world. Every now and then it slips into view – a neatly manicured lawn, shrubs tucked next to aluminum siding, a huge palm tree looking lost amid the bustle of a shopping mall. For many, there is too little nature in the metropolis, and what little there is seems contrived, pathetic. In fact, nature is there, but it has been so thoroughly controlled and mastered that, in a sense, it ceases to exist. Members of the metropolis take the domination of nature for granted. Indeed, the conquest of nature is so central to American culture today that we hardly give the idea a second thought.

A great deal of arrogance surrounds this late-twentieth-century attitude toward the environment, and a great deal of history as well. This book explores the role of the Industrial Revolution in this aggressive stance toward the natural world. The transformation of nature is at least as old as our presence as a species on this planet. But the advent of the industrial age marked a shift in humankind's relations with the earth. I am concerned mainly with describing this shift as it was felt in New England, to journey back to a time when the task of subduing nature was full of hard-fought battles and much less arrogance.

This study began in 1984 as a research paper that I wrote while at Brandeis University. That paper expanded over the years into this book with nourishment from a number of people and institutions. I received generous financial support from the Irving and Rose Crown Fellowship, which I held while at Brandeis, and from the Museum of American Textile History, which granted me a Sullivan Fellowship and offered additional funds to support the publication of this book. The Michigan Society of Fellows at the University of Michigan, Ann Arbor, offered aid during the last stages of this project.

The curatorial staffs at the following institutions provided indispensable research assistance: Baker Library, Harvard Business School, Boston Public Library, Charles River Museum of Industry, Charles River Watershed Association, Dedham Historical Society, Manchester Historic Association, Massachusetts Archives at Columbia Point, Massachusetts Historical Society, Merrimack River Watershed Council, Museum of American Textile History, New Hampshire Historical Society, New Hampshire State Library, New Hampshire Supreme Court Library, and the University of Lowell.

My greatest debt is to my dissertation advisers: David Fischer, Morton Horwitz, and Donald Worster. For most historians, these names need no introduction. Together they taught me how to be a historian. Beyond that, they pushed me hard to think critically, write effectively, and speak my mind. And through it all, they showed me more patience than I perhaps deserved. I am proud to call them my teachers.

No one has endured more discussions about this book than Michael Clark. His warmth, unfailing support, and intellectual integrity have made this a vastly more successful, ambitious, and rewarding endeavor. Robert Cohen has been my dear friend and an incisive critic of my writing for some years now. Bruce Mizrach helped with that sharp mind of his, bringing nuance and subtlety to this work. Robert Hannigan and Jim O'Brien read the entire manuscript while it was still a dissertation and contributed invaluable insights and suggestions. Without Wynn Schwartz's sensitivity, without his kindness, his dedication, and most of all, without his interpretations, these last several years and this book would not have been the same.

I am also grateful to John Demos, Brian Donahue, Paul Faler, Thelma Fleishman, Michael Folsom, Richard Galdston, Gerald Gill, Hayes Gladstone, Kerstin Gorham, Paul Karon, Morton Keller, James Kloppenberg, Marcel La Follette, Arthur McEvoy, Gail Fowler Mohanty, Jeffrey Seidman, Deborah Steinberger, Jeffrey Stine, Joel Tarr, and Reed Ueda for the friendly advice and encouragement that they have offered over the years.

I enjoyed working with Frank Smith, who shepherded the book through its final stages. My thanks to Robert Racine and Mary Racine for their help with the editing of the manuscript. David Forman drew the maps with an eye for detail that I greatly admire. Parts of Chapter 4 first appeared in "Dam-Breaking in the 19th-Century Merrimack Valley: Water, Social Conflict, and the Waltham–Lowell Mills," *Journal of Social History* (Fall 1990): 25–45.

My father never knew about this book, his life cut short before it even started. I like to think, however, that the values my mother and he instilled, the creativity they encouraged, and the honesty they expected fueled this enterprise.

Maria Del Monaco was always there for me. Always.

ABBREVIATIONS

AMC	Amoskeag Manufacturing Company
BCRD	Belknap County Registry of Deeds, Laconia, N.H.
BL	Baker Library, Harvard Business School, Boston
BMC	Boston Manufacturing Company
BSNH	Boston Society of Natural History
CDL	*City Documents of the City of Lowell*
EC	Essex Company
JMC	Jackson Manufacturing Company
Lake Co.	Winnepissiogee Lake Cotton and Woolen Manufacturing Company
LET	*Lawrence Evening Tribune*
LVT	*Lake Village Times*
Mass. Archive	Massachusetts Archives at Columbia Point, Boston
MATH	Museum of American Textile History, North Andover, Mass.
MHA	Manchester Historic Association, Manchester, N.H.
MHS	Massachusetts Historical Society, Boston
MMC	Merrimack Manufacturing Company
MSBH	Massachusetts State Board of Health
NHSCN	New Hampshire Supreme Court Notes, New Hampshire Supreme Court Library, Concord
NMC	Nashua Manufacturing Company
ORHA	*Contributions of the Old Residents' Historical Association, Lowell, Mass.*
PLC	Proprietors of Locks and Canals on Merrimack River
SJC	Supreme Judicial Court

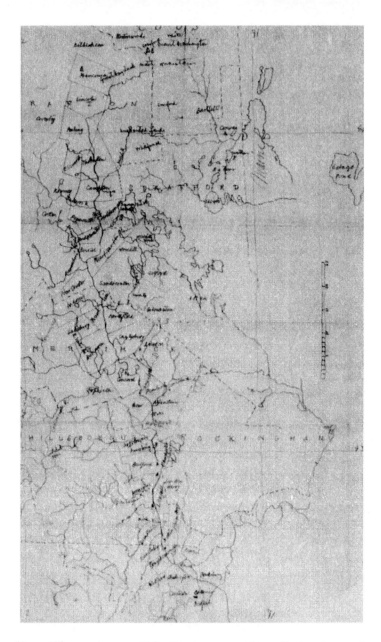

Henry Thoreau's map of his 1839 journey with small triangles marking where he and his brother camped overnight. Henry David Thoreau Collection (no. 6345). (Courtesy of the Clifton Waller Barrett
Library, Manuscripts Division, Special Collections Dept., University
of Virginia Library.)

INTRODUCTION

In the summer of 1839, Henry David Thoreau and his brother
John prepared to leave Concord, Massachusetts, on a two-week
holiday. Earlier that spring, Henry had built a fifteen-foot boat
– flat bottomed, narrow, with a sharp prow – which he painted
blue and green and christened the *Musketaquid*, the Indian
name for the Concord River. The Thoreau brothers loaded the
boat with their camping gear: a lantern, kettle, rice, sugar,
bread, cocoa, and some potatoes and melons they had grown
themselves. They packed a cotton cloth to use as a tent or sail,
two buffalo skins for sleeping, and an extra blanket. They also
brought along a gazetteer to guide them in their wanderings, to
flesh out the landscape that would soon unfold before them. As
dreary morning gave way to mild afternoon, they launched
their boat into the Concord River and began what turned out to
be a special journey, a trip that took them through New England and brought them close to the soul of the land.[1]
  The Thoreau brothers headed for the White Mountains, a
destination that soon led them up the Merrimack River. At the
time, the Merrimack valley was in the midst of industrial
change. As they journeyed, Henry Thoreau diligently searched
for the peace and quiet of the wilderness. But the unmistakable
signs of industrial transformation – the factories, cities, and
railroads – intruded on his plans. This was not the path he
cared to see New England take, and much of what he saw along
the way disturbed him. What did the region look like when
Henry Thoreau made his journey there?[2]

---

[1] Henry D. Thoreau, *The Illustrated "A Week on the Concord and Merrimack Rivers"*
(Princeton, N.J., 1983), xx, 15 (references to the gazetteer appear on pp. 115, 196,
and 245); William Howarth, ed., *Thoreau in the Mountains: Writings by Henry David
Thoreau* (New York, 1982), 206; Henry D. Thoreau, *Journal*, ed. John C. Broderick et
al., 2 vols. (Princeton, N.J., 1981), 1:124–5.
[2] The following description is based on contemporary accounts of the land and water-

When they left home, the Thoreau brothers ventured about seven miles down the Concord River past its marshes and meadows. They stopped in Billerica, where they found a small piece of rising ground and pitched their tent for the night.[3] The next day, Henry and John left the Concord River, passing through the Middlesex Canal into the Merrimack River. They rushed through the canal in less than an hour. After spending a day lolling in the Concord, enjoying its broad expanse of meadows and its button bushes off to the side, the Middlesex Canal – built by practical-minded Yankees and so clearly artificial – annoyed Henry Thoreau.[4]

It struck Thoreau how easily they passed from the Concord to the Merrimack: "It seemed a strange phenomenon to us that the two rivers should mingle their waters so readily, since we had never associated them in our thoughts."[5] The two rivers indeed offered a stark contrast. Compared with the sedate waters of the Concord River, which Thoreau described as a "dead stream,"[6] the Merrimack's waters were far more powerful, tumbling at one point more than fifty feet in half a mile. By 1839, industry had reared itself on the Merrimack River as it never would on the Concord. A glance east from the Middlesex Canal proved the point. Perched along the shore of the Merrimack, and dependent on the river for its livelihood, was a flourishing city of about twenty thousand people. Here the river bent sharply north, then south, with a level plain set in the bend – a perfect surface for a city.[7]

Brick buildings hovered over a network of canals that cut the land into angular patches. Arranged neatly in clusters, the buildings looked nearly the same, a sea of uniformity broken occasionally by the bell towers that marked off one complex from the next. Everything here seemed set in motion, the air thick with progress. And water, the flow of the Merrimack, was at the heart of it all.

Water was forced into the canals by a masonry and wood dam that spanned one thousand feet across the Merrimack River.[8]

---

scape, company records, Thoreau's own descriptions, and secondary sources that describe the physical features of the Merrimack valley.

[3] Thoreau, *Journal*, 1:134.

[4] Thoreau, *A Week*, 62.

[5] Ibid., 80.

[6] Ibid., 61.

[7] Margaret T. Parker, *Lowell: A Study of Industrial Development* (New York, 1940), 66.

[8] Peter M. Molloy, ed., *The Lower Merrimack River Valley: An Inventory of Historic Engineering and Industrial Sites* (Washington, D.C., 1976), 80–1.

As the water traveled through this complicated system of canals, it was at times diverted into the basements of the brick mill buildings. Concealed below ground were large wheels with buckets attached to their rims; as water filled each bucket, the weight made the wheels turn slowly.[9] The wheels were connected to a system of belts, shafts, and pulleys that dispatched power up several floors and across their ceilings. Belts hanging from the ceiling were linked to machinery that spun cotton yarn and wove cloth from the yarn. The process was simplicity itself: water, gravity, and then power and production.

Twenty-eight mills worked here on several hundred acres of land. Six days a week, the noise of about 150,000 spindles and close to 5,000 looms could be heard across the land. The mills were driven by a massive hydraulic apparatus designed with the single-minded purpose of controlling water for production. The factories employed about eight thousand people – three-fourths of whom were women – and produced roughly 50 million yards of cloth each year, including sheetings, calicoes, broadcloths, carpets, and rugs.[10] This was hardworking, calculating Lowell, Massachusetts – and this was a place Thoreau and his brother chose not to visit. Instead, after leaving the Middlesex Canal, the two pointed their boat upstream, away from the city. They rowed hard against the river's current and, in a symbolic way, against the social currents of the time.[11] But Thoreau, appalled by the progress represented by Lowell, would be further disappointed as he made his way up the river.

Upstream the river looked wide and the banks steep, features conducive to storing the water that drove the mills below. Above and beyond the banks lay the small hills and fertile terrain of Dracut and Chelmsford, both of deeply rural character. Although industry had settled on the Merrimack, vast stretches of the valley were still devoted to agriculture. But the signs of industry were never far off. It was not happenstance that caused the level of the water to appear rather high up the banks at this point. Indeed, the two brothers rowed in water that was part of an eighteen-mile stretch raised by the dam at Lowell.

9 Patrick M. Malone, *Canals and Industry: Engineering in Lowell, 1821–1880* (Lowell, 1983), 3, 5.
10 John Hayward, *The New England Gazetteer Containing Descriptions of All the States, Counties and Towns in New England,* 2d ed. (Concord, N.H., 1839), unpaginated, see under Lowell, Mass.
11 This point is made in Linck C. Johnson, "A Natural Harvest: The Writing of *A Week on the Concord and Merrimack Rivers,* with the Text of the First Draft" (Ph.D. diss., Princeton University, 1974), 231.

Where once a canal was needed to pass by the Wicasee Falls in Tyngsborough, now a level expanse of water had buried the falls.[12]

The dam at Lowell did more than raise the river's water. Examining the Merrimack upstream from the dam, Thoreau contemplated progress's price: "Shad and alewives are taken here in their season, but salmon, though at one time more numerous than shad, are now more rare. Bass, also, are taken occasionally; but locks and dams have proved more or less destructive to the fisheries."[13]

That night they camped under oak trees in Tyngsborough at a point where the Merrimack broadened out, making the water appear placid, almost listless.[14] They made a fire and ate their dinner. Drifting off to sleep, Henry Thoreau recalled how far they had come, from the serene wilds of Concord River to the bustling Merrimack: "Instead of the Scythian vastness of the Billerica night, and its wild musical sounds, we were kept awake by the boisterous sport of some Irish laborers on the railroad, wafted to us over the water, still unwearied and unresting on this seventh day."[15]

They awoke the next morning and, leaving Massachusetts behind, came into New Hampshire. They approached the village of Nashua, built at the confluence of the Nashua and Merrimack rivers. In 1839, the village had roughly six thousand people and two textile companies operating five cotton mills.[16] The mill buildings were virtually identical to those found in Lowell, about 150 feet long, 45 feet wide, and four to six stories high.[17] A canal three miles long, 35 feet wide, and 6.5 feet deep carried water to the mills of the Nashua Manufacturing Company, where it fell roughly 30 feet.[18] Massive waterwheels, 30 feet in diameter, received the water, the lifeblood of the facto-

---

[12] John Warner Barber, *Historical Collections, Being a General Collection of Interesting Facts, Traditions, Biographical Sketches, Anecdotes, & c. relating to the History of Antiquities of Every Town in Massachusetts, with Geographical Descriptions* (Worcester, Mass., 1839), 375, 386; Molloy, *Lower Merrimack River Valley*, 81; Mary Stetson Clarke, *The Old Middlesex Canal* (Melrose, Mass., 1974), 106.

[13] Thoreau, *A Week*, 88.

[14] Thoreau, *Journal*, 1:136.

[15] Thoreau, *A Week*, 115–16.

[16] U.S. Census of Population, 1840.

[17] Hayward, *New England Gazetteer*, see under Nashua, N.H.

[18] Treasurer's and Annual Reports, 1825–56, AB-1, NMC Papers, BL. Hayward's *New England Gazetteer* offers conflicting figures for canal width and depth of sixty and eight feet, respectively.

ries.[19] The Jackson Manufacturing Company's canal was half a mile long with 20 feet of water available to drive its two mills.[20] Seeing the mouth of the Nashua "obstructed" by these factories, Thoreau and his brother did not dally there. Instead, they rowed on as the sun began to set, looking for a place to spend the night. They settled on a site near Penichook [Pennichuck] Brook, a short stream flowing into the Merrimack from the west, and spread out beneath some pine trees.[21]

A thick fog enveloped them the next morning. Passing the mouth of Penichook Brook without noticing it, they squeezed upstream between the towns of Merrimack and Litchfield. No villages were in sight, just woodland and pasture interspersed with occasional fields growing corn, rye, or maybe some English grass.[22] Sometime before noon, they ascended Cromwell's Canal, a short canal with only one lock, having waited for several other boats to pass through.[23] These boats might have been carrying cords of wood, bricks, or potash, common items sent downstream, through the Middlesex Canal, and on to Boston. Boats sailing upstream hauled cotton, grown in the South and shipped to Boston, machine parts, and even entire textile machines, some so wide that mill agents worried whether they would fit through the canals' locks.[24]

A while later, they stopped to rest on an island before continuing on to Moore's Falls.[25] At the falls, they entered the Union Canal, a series of seven locks, encompassing six falls in the course of nine miles.[26] As they disappeared upstream, the bustle of Nashua and Lowell seemed more remote than ever. Before sunset they entered the canal at Coos Falls and stopped for the night in Bedford, on the river's west side.[27] There a rural expanse appeared before them, a reminder of agriculture's continued place in the valley.[28]

19 Treasurer's and Annual Reports, 1825–56.
20 Hayward, *New England Gazetteer*, see under Nashua, N.H.
21 Thoreau, *A Week*, 162, 171.
22 Ibid., 192, 194.
23 Ibid., 196, 200; Christopher Roberts, *The Middlesex Canal: 1793–1860* (Cambridge, Mass., 1938), 129.
24 The letters of Oliver Dean, agent of the AMC in Manchester, N.H., describe what was carried on the river. See Oliver Dean, Outgoing Correspondence, 1826–31, box 1, vol. 1, AMC Papers, MHA.
25 Thoreau, *A Week*, 222.
26 Ibid., 231; Roberts, *Middlesex Canal*, 129.
27 Thoreau, *A Week*, 234; idem, *Journal*, 1:136.
28 Hayward, *New England Gazetteer*, see under Bedford, N.H.

Proceeding upstream, the Thoreau brothers rowed hard the next day past Cohas Brook and then surmounted Goff's, Short, Griffin's, and Merrill's falls. Ahead the river dropped a sharp fifty-four feet within half a mile, while several islands of varying dimensions sat unperturbed below the falls. The falls here were called Amoskeag, after the Indian word for fishing place.[29] Large rocks scattered about the river obstructed the passage of fish, making this the most noted spring fishery on the river. Fishermen laid claim to these rocks, leaning off them to snag fish in the water below.[30] We can imagine the summer sun beating down on the rocks when the Thoreau brothers arrived. They were probably deserted after the end of the spring fishing season, and with the river's stock of fish declining, these rocks would one day be permanently abandoned.

A little further downstream, Henry Thoreau and his brother prepared to enter the Amoskeag Canal around the falls. At the time, the area around Amoskeag Falls was under construction, a sign of impending industrial development. On the east side of the falls, a stone dam, planned to average eight feet in height in one section and thirty-four feet high in another, was being built.[31] Surrounded by the din of workmen building, Henry Thoreau made "haste to get past the village here collected, and out of hearing of the hammer which was laying the foundation of another Lowell on the banks."[32]

The new industrial city of Manchester was being built here. When the two brothers arrived in the summer of 1839, the Amoskeag Manufacturing Company, owner of much of the land in sight, had completed a power canal about a mile in length and was working on another in order to operate a system of water control much like the one at Lowell. It had one cotton mill in partial operation and another one nearly finished. It had laid out streets, mill yards, and boardinghouses, and had nearly completed a dam across the river.[33] From such modest begin-

---

[29] Dena F. Dincauze, *The Neville Site: 8,000 Years at Amoskeag, Manchester, New Hampshire* (Cambridge, Mass., 1976), 1.

[30] George E. Burnham, "Amoskeag's Old Fishing Rocks," *Manchester Historic Association Collections* 4 (1908): 60–7; Myron Gordon and Philip M. Marston, "Early Fishing Along the Merrimack," *New England Naturalist*, Sept. 1940, 3–4.

[31] The figures are from Directors' Records, 4 Mar. 1837, vol. A-2, AMC Papers, BL. The papers of the AMC are divided between BL and the MHA.

[32] Thoreau, *A Week*, 245.

[33] Treasurer's Report, 1 July 1839, vol. AD-1, AMC Papers, BL.

nings, Manchester would rise to prominence as New Hampshire's most impressive textile city.

The Thoreau brothers meanwhile moved on toward Hooksett, their final destination by water. There, on the east bank, they found a quiet corner of the river to leave their boat and proceeded by foot or stage for the next week.[34]

On 5 September, they walked ten miles to Concord, New Hampshire, and from there went by coach to Plymouth. The two brothers spent the next four days hiking through the White Mountains and visiting its sites before riding back to Concord, New Hampshire, on 11 September 1839. They went to Hooksett the next day, picked up their boat, and set off downstream. The towns must have passed by in a blur: Hooksett, Manchester, Bedford, and finally Merrimack, where they stopped for the night. The following day they retraced their path to the Middlesex Canal and reached their home in Concord, Massachusetts, on 13 September.[35] "We had made about fifty miles this day with sail and oar," wrote Thoreau, "and now, far in the evening, our boat was grating against the bulrushes of its native port, and its keel recognized the Concord mud."[36]

Although their vacation had come to an end, the trip lived on in Henry Thoreau's mind. In 1849, he immortalized the vacation when he published *A Week on the Concord and Merrimack Rivers*. Thoreau did not write a complete account of their voyage. Rather, the book has a rambling, philosophical tone, with much to say about the natural world. Among Thoreau's chief concerns are nature's timeless, eternal qualities. Scattered among the pages are references to nature's enduring features, to its vigor and perseverance despite what its foes, farmers and industrialists alike, had done.[37]

But one wonders about nature's supposed immortal qualities, especially in the face of formidable industrial transformation. As Thoreau struggled to write the book in the 1840s, industry proceeded apace. The landscape he viewed in 1839 had been radically altered by the time the book appeared. Indeed, in the ten years it took Thoreau to publish his thoughts, the textile city of Lawrence, Massachusetts, had gone up along the river's

---

34 Thoreau, *A Week*, 291; idem, *Journal*, 1:136–7.
35 Thoreau, *Journal*, 1:136–7.
36 Thoreau, *A Week*, 393.
37 See, e.g., ibid., 34, 62.

shores as industry continued to change the look of the Merrimack valley. A river that had once flowed triumphant and
unimpeded across New England had slowed to an industrial
pace – a development Thoreau lamented. When Thoreau re-
visited southern New Hampshire nine years after his original
voyage, he noted the extension of the railroad, the rapidly
growing population, and the relentless tide of industry and how
it had transformed the river valley.[38] In Thoreau's words, "Instead of the scream of a fish-hawk scaring the fishes, is heard the
whistle of the steam-engine arousing a country to its progress."[39]

If Thoreau found industry's imprint on New England's landscape unwelcome, he was no more heartened by what the region's farmers had done. In his eyes, Yankee farmers had already played havoc with the land, tromped all over it, fenced it
where they pleased, planted orchards and market gardens, and
ruined a once fertile New England soil. Their pockets interfered
with their better judgment, causing them to treat the land inso-
lently. According to Thoreau, the commercial economy had
made matters worse for farmers. It urged them on a reckless tear
through the land, chopping down trees, plowing up the soil,
reclaiming meadows – all in a desperate attempt to seize whatever economic opportunity the market offered. Even before the
factories arrived, the nineteenth-century New England landscape seemed poisoned to Thoreau, the victim of the unsur-
passed greed of a market-oriented people.

The world, Thoreau believed, had not always been this way.
He consoled himself with thoughts of preindustrial life in
which farmers worked the land to satisfy only themselves and
not the demands of the market. He romanticized life in former times, imagining a world of economic self-sufficiency, of
lives filled with meaningful work and divinity.[40] Life before the
Europeans arrived seemed even better to him. The Indians
who lived on the land respected nature, cherishing a land that
white settlers took for granted. Throughout *A Week,* Thoreau
inquires into the fate of these people, sympathizing with them,

---

[38] Ibid., xvii, 245. The population of Manchester, N.H., had grown from 3,235 in 1840
to 13,932 ten years later. See U.S. Census of Population, 1840, 1850.

[39] Thoreau, *A Week,* 87.

[40] Thoreau's views on agriculture and the economy of New England are discussed in
Robert A. Gross, "The Great Bean Field Hoax: Thoreau and the Agricultural Re-
formers," *Virginia Quarterly Review* 61 (1985): 483–97, esp. 490–1; idem, "'The Most
Estimable Place in All the World': A Debate on Progress in Nineteenth-Century
Concord," *Studies in the American Renaissance* 2 (1978): 1–15.

for they were the antithesis of progress, the epitome of harmonious relations with the New England soil.[41]

Thoreau sought a world long gone in 1839. European colonization had obliterated Native American culture, spreading diseases unknown to the Indian immune system and imposing another set of environmental values and behaviors on the landscape. As the eighteenth century progressed, the marks of European culture became more firmly entrenched on the land: Rye, oats, and English hay were planted, swine and cattle were grazed, and the most visible manifestation of European settlement appeared – the stone fence. A new European ecological pattern had conquered its Indian predecessor. But this too was passing. And some day, the stone fences that set one field off from another would run through new-growth forests, serving as subtle reminders of New England's agricultural past. As forests would encroach on the weed-strewn fields, long since given up, another ecological transformation would begin.[42]

When Thoreau made his first journey through New England, he saw signs of all these environmental patterns – the Indian's occupancy, the white man's agriculture, and the new industrial regime. Agriculture still dominated much of the land as well as the lives of many of its inhabitants. Yet agriculture was being pushed aside at points to accommodate the new urban–industrial stirrings. The pages that follow look at what is perhaps the best-known aspect of New England's industrial transformation: the emergence of the Waltham–Lowell system. Francis Cabot Lowell's Waltham experiment on the Charles River and the later founding of Lowell, Massachusetts, are part of almost every textbook treatment of early American industry. A group of New England entrepreneurs known collectively as the Boston Associates, with the help of a corps of expert mechanics and engineers, were the architects of the Waltham–Lowell mills. They were responsible for a number of technological and labor innovations that made large-scale production of cotton textiles a flourishing industry in nineteenth-century New England.[43]

---

41 Johnson, "Natural Harvest," 222–3, 247–8.
42 William Cronon, *Changes in the Land: Indians, Colonists, and the Ecology of New England* (New York, 1983), 156.
43 For an early account of the rise of the Waltham–Lowell system, see Caroline F. Ware, *The Early New England Cotton Manufacture: A Study in Industrial Beginnings* (Boston, 1931).

This is a familiar story, one told by historians for some time now.[44] Much of the attention has focused on the legendary textile city of Lowell. Early works concentrated on such diverse topics as technological change, industrial architecture, and labor relations. More recent works have explored the intellectual and cultural responses to industrial growth in this great urban center. Some have studied textile and waterpower technology, while others have examined the city's immigrant communities. Business historians have explored the Boston Associates' plans for the founding and design of the city. Issues of gender, class, family, and work, topics of concern to social historians, have also been investigated. In all, the range of subjects and approaches is impressive.[45]

Still, an important part of this story remains untold. Nature has tended to be excluded from the history of industrial transformation. The natural world has been there all along, but historians have largely neglected it. One of the points of this study is to show just how important nature – in this case water – was to the industrial and urban history of New England. History is forever being written and rewritten. And no claims are

[44] The tremendous attention paid to the Waltham–Lowell mills has prompted some to question whether our understanding of industrialization has been skewed. Indeed, most textile mills built in nineteenth-century New England were not based on the Waltham–Lowell model. Most relied on a more diverse set of laborers who found their housing throughout the community, not in boardinghouses common in Lowell. The mills existed on small amounts of capital and were often run by at least some of their owners. They managed to meet their needs with small quantities of waterpower and were commonly found in southern New England. See Jonathan Prude, *The Coming of Industrial Order: Town and Factory Life in Rural Massachusetts, 1810–1860* (Cambridge, 1983), xiv–xv.

[45] See George S. Gibb, *The Saco-Lowell Shops: Textile Machinery Building in New England, 1813–1949* (Cambridge, Mass., 1950); John P. Coolidge, *Mill and Mansion: A Study of Architecture and Society in Lowell, Massachusetts, 1820–1865* (1942; reprint, New York, 1967), and the discussion of labor in Lowell in Norman Ware, *The Industrial Worker, 1840–1860: The Reaction of American Industrial Society to the Advance of the Industrial Revolution* (1924; reprint, Chicago, 1964); Thomas Bender, *Toward an Urban Vision: Ideas and Institutions in Nineteenth Century America* (Baltimore, 1975); John F. Kasson, *Civilizing the Machine: Technology and Republican Values in America, 1776–1900* (New York, 1976); David J. Jeremy, *Transatlantic Industrial Revolution: The Diffusion of Textile Technologies Between Britain and America, 1790–1830s* (Cambridge, Mass., 1981); Louis C. Hunter, *Waterpower in the Century of the Steam Engine,* A History of Industrial Power in the United States, 1780–1930, vol. 1 (Charlottesville, Va., 1979); Brian C. Mitchell, *The Paddy Camps: The Irish of Lowell, 1821–61* (Urbana, Ill., 1988); Robert V. Spalding, "The Boston Mercantile Community and the Promotion of the Textile Industry in New England, 1813–1860" (Ph.D. diss., Yale University, 1963); Robert F. Dalzell, Jr., *Enterprising Elite: The Boston Associates and the World They Made* (Cambridge, Mass., 1987); and Thomas Dublin, *Women at Work: The Transformation of Work and Community in Lowell, Massachusetts, 1826–1860* (New York, 1979).

made here to have documented the definitive story behind industrial change. But a rather different way of understanding industrialization will be offered: an environmental approach.

New England's productive output expanded as new technologies were manipulated and applied to the region's available natural resources. The natural world came to represent new sources of energy and raw materials. Nature was perceived more and more as a set of "inputs" central to the productive capacity of the economy. The broadening of New England's industrial base allowed more products to be manufactured. But this leap in productive capacity also meant a change in how people understood and made use of the environment around them. The burden of this study is to show that industrial capitalism is not only an economic system, but a system of ecological relations as well.

In a sense, the environmental perspective proposed here is a counterpart to the work of social historians. They view industrialization as a transformation in social relations, in people's interactions with one another. But if industrial transformation affected such aspects of social life as class, gender, and family relations, it also altered human relations with the natural world. A close connection exists between the way a culture tends to its natural resources and the way it employs its human resources.[46] In this book, I will look at how people reshaped the natural world, how they transformed their relationship with nature to generate economic growth.

Environmental history is a new field of study. Its central task is twofold: to examine how the environment has shaped human cultures over time and to understand the effects of human activity on nature itself. Very simply, environmental historians want to make sense of the role that the natural world has played in the historical process. How did the natural order affect how history was made? What impact did human-induced changes have on the environment, and how did such changes limit the possible paths that history could take?[47]

A number of approaches and assumptions inform the work

---

[46] See Maurice Godelier, *The Mental and the Material: Thought Economy and Society*, trans. Martin Thom (London, 1986), 121.

[47] For an introduction to the field, see Donald Worster, "History as Natural History: An Essay on Theory and Method," *Pacific Historical Review* 53 (1984): 1–19. The field's historiography is reviewed in Richard White, "American Environmental History: The Development of a New Historical Field," ibid., 54 (1985): 297–335.

now being done in this field. For me, one assumption is absolutely central: Human history is defined by the transformation and control of nature. The progressive reworking of the natural world to suit human needs and desires has been a constant throughout our history on this planet. As the anthropologist John Bennett writes, humankind's relationship with the environment "has featured a growing absorption of the physical environment into the cognitively defined world of human events and actions."[48] To one degree or another, nature has always been incorporated into human cultures. The point is to understand this phenomenon historically. The emergence of industrial capitalism, I will show, marked a new stage in this continuing process, one in which nature has been more thoroughly defined in anthropocentric terms. Industrial expansion involved a profound restructuring of the environment – a far more comprehensive incorporation of nature into the human agenda than ever existed before. At its core, the process entailed a systematic effort to control and master nature, a development that had dramatic implications for both human beings and the environment itself.

New England's rivers and streams, with their abundance of waterpower, provide an important place to observe this historical process. Carving their way through the land, the region's rivers, with their relatively high average slopes and frequent falls, were perfect for America's early industrial development. The Boston Associates poured capital into a long list of New England river valleys, but only two will be singled out for study here. The rise of the Waltham–Lowell system began along the Charles River in Waltham. As the nineteenth century progressed, the ambitions of the Boston Associates outgrew the modest waterpower of the Charles. The search began for more substantial sources of energy, and before long the well-endowed water resources of the Merrimack River were tapped. The textile cities of the Merrimack valley – Lowell, Lawrence, Nashua, and Manchester – controlled water to an unparalleled

---

[48] John W. Bennett, *The Ecological Transition: Cultural Anthropology and Human Adaptation* (New York, 1976), 4. My environmental perspective on history borrows from the fields of economic and ecological anthropology, particularly the contributions of Karl Polanyi, Roy Rappaport, Marvin Harris, Richard Lee, Maurice Godelier, and Marshall Sahlins. On anthropological theory and industrial change, see Theodore L. Steinberg, "An Ecological Perspective on the Origins of Industrialization," *Environmental Review* 10 (1986): 261–76.

degree, and for this reason, the valley will occupy the bulk of this study.

This book has three primary goals. First, it seeks to advance the study of New England's ecological history by approaching industrial capitalism from an environmental perspective. In addition, it focuses on the conflict that developed over nature, in hopes of expanding our understanding of social history in the industrial age. Finally, it strives to contribute to legal history by exploring the relationship between law, water, and economic change.

Environmental historians have already paid a good deal of attention to colonial New England.[49] European settlement and its history in the following centuries altered the landscape of the region, marking it off in segments and patches that honored the abstraction of private property. The colonists saw before them a land sparkling with opportunity, a world filled with natural wealth just waiting, in their minds, to be "improved." Early New Englanders were amazingly successful at reshaping the natural world to meet their own economic needs, at incorporating nature into their own distinct culture. The thrust then of New England's history in the decades leading up to the nineteenth century tended toward the expansion of natural resource use, toward the more thoroughgoing *commodification of nature*. Simply put, this was a process whereby nature – all things and relations in it – was conceived of, acted upon, and valued primarily for its capacity to be exchanged at market for profit.

Unlike the Indians who preceded them, European settlers brought along a culture that viewed the land, and what was on it, as a source of profit. The Europeans tended, in short, to transform nature into discrete bundles of commodities. This is not to say that the colonists saw only personal gain in the landscape. But when compared to their Indian predecessors, the colonists tended to integrate the natural world into the money economy, into capitalist economic relations. To New England's white settlers, then, the region's fish, fur, and timber often represented commercial opportunity. By the nineteenth cen-

---

[49] The two seminal works here are Cronon, *Changes in the Land*, esp. pp. 54–81, which discusses changing property relations, and Carolyn Merchant, *Ecological Revolutions: Nature, Gender, and Science in New England* (Chapel Hill, N.C., 1989). Merchant's book also considers the nineteenth century.

tury, this early version of capitalism – with its distinct environmental results – was evolving into a more advanced stage. It retained a commitment to markets and private property, but increasingly took the form of class relations and wage labor. This maturation also had environmental import. The relevant question for this study is how rivers and streams fared as capitalism matured into its modern, corporate, wage-based form. If capitalism entailed the commodification of nature, how did that commodification change, if at all, as capitalism itself changed?

The colonists were largely successful in sectioning off the land, in setting up stone walls to distinguish one person's property from another. Rivers and lakes, however, could not be fenced off. The New England waterscape, like the waters of the rest of the world, resisted attempts to pin ownership on it. Water, in short, is more difficult to commodify and privatize than land. William Blackstone, the eighteenth-century jurist, believed that a stream's ever-changing physical nature often prevented the recognition of explicit property rights. Water, he explained, "is a moveable, wandering thing, and must of necessity continue common by the law of nature; so that I can only have a temporary, transient, usufructuary property therein."[50] By its very nature, water is a common resource – a part of the natural world not easily subject to private ownership. For a society wedded to the institution of private property, water raises troubling questions of control, ownership, and regulation.[51]

Although the colonists had used water for power – to grind grain, saw logs, and full cloth – the history of water use in the nineteenth century marks a break with this past. The advent of large, integrated textile factories with substantial demands for energy transformed the way New England's waters were used.

[50] William Blackstone, *Commentaries on the Laws of England*, 4 vols. (1783; reprint, New York, 1978), 2:18.

[51] A vast theoretical literature exists on common resources. For a useful introduction to the subject, see H. Scott Gordon, "The Economic Theory of a Common-Property Resource: The Fishery," *Journal of Political Economy* 62 (1954): 124–42. The classic work on this issue is Garrett Hardin, "The Tragedy of the Commons," *Science* 162 (1968): 1243–8. Critiques of Hardin's essay can be found in Garrett Hardin and John Baden, eds., *Managing the Commons* (San Francisco, 1977). Also see Partha Dasgupta, *The Control of Resources* (Cambridge, Mass., 1982), 13–14, 31–7; Carol Rose, "The Comedy of the Commons: Custom, Commerce, and Inherently Public Property," *University of Chicago Law Review* 53 (1986): 711–81; and Arthur F. McEvoy, *The Fisherman's Problem: Ecology and Law in the California Fisheries, 1850–1980* (Cambridge, 1986).

Much of this change forced water to the center of the production process on a new and unheard of scale. The minute, fine-tuned control of water became the hallmark of the Waltham–Lowell mills as they formed and matured throughout New England. The economic development of water, the construction of dams and power canals, created greater quantities of waterpower. And more waterpower meant more water available to be owned, regulated, and sold in the market.

The emergence of industrial capitalism signaled the beginning of a new chapter in New England's ecological history. As the region moved forth into the industrial age, there were all kinds of transformations happening – changes in the workplace, the home, the family, and the church. But few of these changes were as blatant as the effect industrialization had on the environment. The region's river ecology, the natural flow of its streams, its migrating species of fish, all felt the effects of industrial change. Industrialization also produced a radically different human ecology – changes in population density, mortality, water quality, and a new disease environment that all urban dwellers would have to face.

For some time now, historians have described the tension and disruption that accompanied this new mode of production, especially the struggle over the workplace.[52] Conflict can certainly take place in the factory, as workers and owners disagree over wages, hours, or regulations. For many people, their picture of industrial change is precisely this – of strikes and walkouts, of class conflict under the factory roof. That is a legitimate and important understanding. But it seems at times to overlook the more far-reaching effects of industrial transformation. Industrial capitalism is too often conceived in terms of its impact on local communities where factories emerged to transform the fabric of traditional life. It is more properly understood as a phenomenon with enormous geographic reach, with the potential to touch people and places far removed from the actual site of production.

Not only the conflict over the workplace, over wages and hours, but the struggle to control and dominate nature is central to industrialization. The face-to-face relations of power in the factory should be supplemented with a broader vision of

---

[52] For New England, see e.g., Prude, *Coming of Industrial Order*, and Paul G. Faler, *Mechanics and Manufacturers in the Early Industrial Revolution: Lynn, Massachusetts, 1780–1860* (Albany, N.Y., 1981).

conflict going on outside the factory walls. That struggle, at least in part, is over who will control the natural world and to what ends. Industrial capitalism is as much a battle over nature as it is over work, as likely to result in strife involving water or land as wages or hours.

New England ventured down the industrial path, but what a rocky, contested path it was, with conflict over water continually plaguing those who sought control of the region's vast waters. Who was to gain title to the water, who were to be the winners and who the losers? That was a question confronting jurists throughout nineteenth-century New England. And it is an important issue for this study as well. The history of water law will play a central role in this book. The legal rules of water structured the way people responded and behaved toward this resource. Over the course of the nineteenth century, water law developed in a way that suited the growing needs of New England's industrial economy. By midcentury, the law had moved firmly toward an instrumental conception of water use, toward a vision of water and law that sanctioned the maximization of economic growth.

The law provided the framework for settling conflicts over water. But that framework was evolving in a way that reduced water, more and more, to an abstract commodity. Why did the law of water change in this way? The search for an answer takes us beyond the confines of the courtroom, to the broader culture of nineteenth-century New England. As the century progressed, a consensus emerged on the need to exploit and manipulate water for economic gain. A stunning cultural transformation was taking place, a shift in people's very perception of nature. By the latter part of the nineteenth century, it was commonly assumed, even expected, that water should be tapped, controlled, and dominated in the name of progress – a view clearly reflected in the law.

This book is organized along narrative lines that, broadly speaking, chart the rise and decline of the Waltham–Lowell system. Chapter 1 explores both the preindustrial history of the Charles River and the later attempt by the Boston Associates to build factories along its shores. The remainder of the book concerns the expansion of the Waltham–Lowell system in the Merrimack valley.

Chapters 2 and 3 explore the emergence of the textile cities that made the Merrimack valley famous. In Chapters 4 and 5, the emphasis shifts to New Hampshire as the Associates expanded their control over water resources in that state. By mid-century, the Waltham–Lowell system had cast itself with a vengeance across the Merrimack valley's waterscape. Together, these chapters examine both the social and legal conflict that ensued from this development. The following two chapters examine the environmental consequences of industrial transformation, the depletion of fisheries and the growing problems with water quality and human health. The final chapter turns to the history of conflict over water as the Waltham–Lowell system declined.

What the Boston Associates and others had done to the waters of New England deeply troubled Henry Thoreau. He could hardly have enjoyed his 1839 vacation, marred as it was by the sights and sounds of vigorous industrial expansion. How did the waterscape Thoreau saw come to be? How did it evolve after his trip? It is with such questions that we begin our own voyage along the waters of New England.

Part I

ORIGINS

1

THE TRANSFORMATION
OF WATER

In 1605, Samuel de Champlain, one of the first Europeans to
see the Charles River, observed "a very broad river" which
seemed to him to stretch boundlessly westward.[1] In fact, the
Charles began just 26 miles southwest of Boston, but because of
its remarkably circuitous route it traveled over 80 river miles
before reaching Boston Harbor (Map 1.1). The river Champlain saw fell roughly 350 feet as it charted a largely uninterrupted course through the 300-square-mile watershed. In its
first 20 miles, the Charles quickly descended 220 feet to Popu-
latic Pond. From this point, the river sloped gently for 40 miles,
falling 1 foot on average for every mile it flowed, winding its
way back and forth across the land searching for the ocean. At
Newton Upper and Lower Falls it tumbled over 40 feet in
roughly 2 miles past extensive bedrock, dipped another 20 feet
at Waltham, and pushed its way east before entering the Boston
Basin. And then at last there was the sea.[2]

Two centuries after Champlain, farmland stretched out
across the Charles River valley punctuated by houses, barns,
wood lots, mills, and other signs of rural life. Water played an
important role in the lives of the people who lived here. The
valley's waters met the needs of the agrarian economy for food
and fodder, while water-driven mills dotted the shores of the
Charles River. As the river descended from its headwaters in
1812, it faced the task of supplying twenty-three mills with
waterpower. Seven of these were traditional mills that ground
grain, sawed wood, and fulled cloth. The remainder were a
potpourri of industry, a ragtag conglomeration of mills produc-

---

[1] Samuel De Champlain, *Voyages of Samuel De Champlain, 1604–1618*, ed. W. L. Grant
(New York, 1907), 67–8.
[2] The description of the Charles River is based on U.S. Army Corps of Engineers, New
England Division, "Charles River, Massachusetts: Main Report & Attachments" (Waltham, Mass., 1972), 9, 14–15, 68–9, and diagrams opposite p. 68.

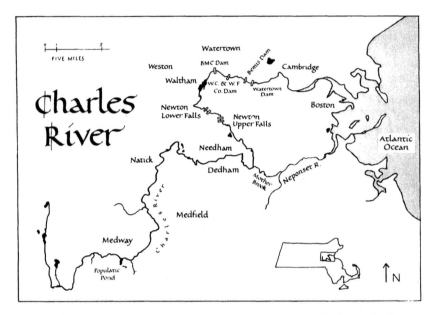

Map 1.1. The Charles River showing dams built along the lower river and the Mother Brook diversion into the Neponset River. (Drawing by David R. Forman.)

ing chocolate, drugs, dyes, nails, paper, screws, snuff, wire, and textiles.[3] This river looked vastly different from the one Champlain saw in the early seventeenth century. The river still traveled the same 300 square miles of basin, it ran the same 80 river miles to its outlet, but the waterscape it traversed looked only vaguely reminiscent of Champlain's day. Progress had reshaped the valley. The water did not course freely throughout the land as it once had but moved, instead, to a more industrious rhythm. Sometimes, especially during dry summer months, it barely moved at all, spreading out across adjacent meadows to await permission from milldam owners to proceed downstream.

This was the valley Patrick Tracy Jackson, Boston merchant and entrepreneur, saw when he came to Waltham, Massachusetts, in 1813 to purchase the land and mill privilege for the Boston Manufacturing Company (BMC). While no single development alone accounts for the transformation of New En-

---

[3] Edward Pierce Hamilton, "Early Industry of the Neponset and the Charles," *Proceedings of the Massachusetts Historical Society* 71 (1953–7): 111.

gland's waters, the incorporation of the BMC in 1813 was a fundamental event along the road to industrial capitalism. A year after its founding by a small group of Boston entrepreneurs, this company erected along the shores of the Charles River in Waltham a brick mill forty-five feet wide, ninety feet long, and four stories high to produce heavy, unbleached, cotton sheeting. This was not, as many believe, the first cotton mill in America to centralize all the processes of production, from carding through weaving, under one roof. But it was the first factory in America to use waterpower to perform all the major steps in making cotton cloth.[4] This innovation formed the basis for the extensive development of New England's waters over the course of the nineteenth century.[5] The BMC was enormously successful, and its Waltham experiment inspired future investment beside far more powerful waters.

What follows is an attempt to assess water use along the Charles River from Champlain's time through the arrival of the BMC. Over the long period encompassed by this chapter, a transformation in the perception and use of water occurred. The crucial period of change happened late in the eighteenth century as cultural norms and legal rules evolved away from a vision of water consistent with long-standing, agricultural uses. Instead, there emerged a different and far more instrumental conception of water centered primarily on economic growth — an understanding of water embraced and furthered by the BMC.

* * *

[4] Kenneth Frank Mailloux, "The Boston Manufacturing Company of Waltham, Massachusetts, 1813–1848: The First Modern Factory in America" (Ph.D. diss., Boston University, 1957), 89.

[5] Historians have discovered that hand power, horsepower, and waterpower were far more central to the early stages of industrial development than steam and coal. In Britain, which had passed through at least two decades of rapid industrial change by 1800, the majority of manufacturing was still carried on without the aid of steam. And where mechanization had taken place, waterpower was used. Water remained a primary source of power in America into the middle of the nineteenth century, well after the steam engine had been invented. As late as 1900, 20 percent of American firms still relied on falling water for their energy. Until the latter part of the nineteenth century, the factories organized on the Waltham–Lowell model depended almost exclusively on water to drive their mills. See Dolores Greenberg, "Reassessing the Power Patterns of the Industrial Revolution: An Anglo-American Comparison," *American Historical Review* 87 (1982): 1237–61; A. E. Musson, "Industrial Motive Power in the United Kingdom, 1800–70," *Economic History Review* 29 (1976): 416–17; George Rogers Taylor, *The Transportation Revolution: 1815–1860* (New York, 1951), 223–4; Jeremy Atack, "Fact in Fiction? The Relative Costs of Steam and Water Power: A Simulation Approach," *Explorations in Economic History* 16 (1979): 430.

Long before water was used on a large scale by industry, New Englanders made use of the natural world around them. The historic cultures of the region – the Indians and the Europeans who followed – related to the land and water in their own distinct ways, but both survived, more or less, by transforming the environment to meet their needs.

There was no escaping the water that flowed throughout New England, and for the Native Americans who first settled here, there was little reason to do so. Much of New England's aboriginal population, according to anthropologist Dean Snow, had a riverine orientation.[6] Particularly for the Eastern Algonquian, a culture that extended from Nova Scotia south to North Carolina, rivers were at the nucleus of local communities with drainage basins tending to define the limits of individual tribal territories. The family units of a number of tribes – and in New England, the Micmac, Malecite, and Penobscot are good examples – located beside streams and ponds.[7] Rivers thus coursed through the center of many aboriginal territories in New England and were put to good use. They often served as arteries of communication and, perhaps more important, as a vital source of food. On both the Charles and Merrimack rivers, archaeological evidence suggests the supply of fish attracted Indian settlements to their shores.[8] The size of individual sites along the Charles, in particular, may well have depended on the quantity of fish available.[9] Many of the Indian communities in the region tended to organize within watersheds, to see their territory as marked off and limited by a single drainage basin.

In contrast to the Indians, white settlers used rivers at first to mark the periphery and limits of their land.[10] When the Massachusetts Bay Company's charter was issued in 1628, the bounds of the grant stretched "betweene a greate river there comonlie called Monomack, alias Merriemack, and a certen other river there called Charles river."[11] At the time the colo-

---

6   Dean R. Snow, *The Archaeology of New England* (New York, 1980), 2–5.
7   Dean R. Snow, "Wabanaki 'Family Hunting Territories,'" *American Anthropologist* 70 (1968): 1143, 1146–8.
8   Dena F. Dincauze, "A Preliminary Report of the Charles River Archaeological Survey (ms. prepared for the National Park Service, 1 Oct. 1968), 17, 27–8, 30; Victoria Bunker Kenyon and Patricia F. McDowell, "Environmental Setting of Merrimack River Valley Prehistoric Sites," *Man in the Northeast* 25 (1983): 13–20.
9   Dincauze, "Charles River Archaeological Survey," 30.
10 Snow, *Archaeology of New England*, 2.
11 Nathaniel B. Shurtleff, ed., *Records of the Governor and Company of the Massachusetts Bay in New England (1628–86)*, 5 vols. (Boston, 1853–4), 1:4.

nists arrived, of course, the landscape had yet to be surveyed in any formal sense, and rivers and streams offered a convenient way to divide up the land. But a great deal more separated the Indians from the white settlers. Two entirely different conceptions of property guided their relations with New England's environment.[12] The Indians tended to have a far more flexible understanding of property rights than the Europeans. New England's Indians hunted, fished, gathered berries, and planted crops to survive. Yet they never owned the land they harvested in any definitive sense, nor the water for that matter. No exclusive property rights inhered. What Indians claimed was the right to whatever lay on that land or in the water. There were no formal hunting and fishing rights; there were, instead, mutual claims to fish and game with several villages overlapping in their use of the land. To be sure, the Indians tapped the region's ecological wealth. But they were guided all along by a nonexclusive conception of property.

The colonists arrived and began to fence in the landscape in a way the Indians never had, hauling stones about and organizing them in lines. Indeed, the stone fence is perhaps the most visible indication that the white settlers had brought with them a rather different vision of property than the one used by the Indians. Fences expressed the colonists' deep commitment to *private* property. The Indians made a living from the land by hunting and gathering and practicing a shifting agriculture, activities consistent with their particular understanding of property. The colonists, in contrast, raised crops and domesticated animals – sheep, swine, and cattle – within specified and fixed property boundaries. They owned the land in a way unknown to their Indian predecessors. For the colonists, exclusive, private rights to the land were recorded in deeds and thus fixed in law. The stone fences that went trailing off across the landscape suggested their dedication to a legal abstraction, to the precept of private property.

With the British system of property relations as their guide, the colonists soon came to see the New England landscape as a vast bundle of commodities – of wood, fish, and furs. In 1632, the British traveler Thomas Morton reflected on New England's fabulous supply of fish, believing the region's cod "a

---

[12] This paragraph and the next are based on William Cronon, *Changes in the Land: Indians, Colonists, and the Ecology of New England* (New York, 1983), 54–81, 127–56.

commodity better than the golden mines of the Spanish Indies." Morton made up a list, what he called a "Catalogue of Commodities," and included on it were some of the most abundant fish that thrived in New England's waters – bass, mackerel, sturgeon, and salmon.[13] Europeans like Morton and others who followed looked at the region's land and water and saw tremendous opportunity before them. They planned to take hold of nature – to till the land, chop down the timber, fish the waters – to survive in this new world. But to the colonists, nature meant more than just a means to survive. To many of them, nature also represented a way to make profits. It was the colonists' intention to own the land and all that was on it, and, at times, to exchange the fruits of that land – its wood, fish, crops, if not the land itself – for money and profit.[14]

The colonists who arrived in the seventeenth century were thus struck by the Charles River's incredible abundance of animal life. Shad, alewives, cod, bass, and smelts filled the river. Along its marsh and shore the colonists found eels, mussels, clams, and, above all, oysters, which the English traveler William Wood reported were "so big that it must admit of a devision before you can well get it into your mouth." Spring brought spawning runs of alewives, "in such multitudes as is allmost incredible, pressing up such shallow waters as will scarce permit them to swimme."[15] Years before Wood wrote these words, between 1617 and 1619, a lethal epidemic of the plague decimated the Massachusetts Indians. The epidemic, which was especially violent around Boston Harbor, probably relieved the pressure on fish stocks in the region's rivers.[16] As European settlers arrived throughout the 1630s, the population of fish in the Charles may well have been in the middle of a fantastic renewal caused by the depopulation of the native people. In any case, the colonists soon partook of this rich resource, which they used for food and fertilizer.

Apart from fish, freshwater meadows found along parts of the Charles and other rivers throughout New England attracted settlers. On these meadows, alluvial grasses – at first

---

[13] Thomas Morton, *New English Canaan*, ed. Charles Francis Adams (1632; reprint, Boston, 1883), 221–8.

[14] Cronon, *Changes in the Land*, 75–6.

[15] William Wood, *Wood's New-England's Prospect* (1634; reprint, Boston, 1865), 38–9.

[16] On the epidemic, see S. F. Cook, *The Indian Population of New England in the Seventeenth Century* (Berkeley, 1976), 29–33; idem, "The Significance of Disease in the Extinction of the New England Indians," *Human Biology* 45 (1973): 487, 489.

wild species and later a mixture of wild and domestic types — grew every year, naturally, without planting and cultivation. The grass was a ready source of hay that farmers used as fodder for their livestock. And by keeping their cattle and other animals well fed, farmers obtained manure for sustaining the fertility of their farmland. Early settlers planned their communities near freshwater meadows, and hay from them soon became an important link in the region's agriculture.[17] The towns of Concord and Sudbury, Massachusetts, were founded in 1635 and 1638, respectively, near river meadows. Similar meadows in the Mystic and Nashua river valleys lured settlers to points along the nearby shores. On the Charles, colonists settled Dedham in 1636 and later Medfield further upstream at those spots on the river where meadow grass flourished.[18]

Farmers dug trenches in the meadows to drain them, providing patches of firm ground for teams of workers to stand on while they harvested the hay. The meadows generally remained flooded during the winter and spring. In the summer, especially during late July and early August when the water flowed out of the meadows, the hay was cut. If nature cooperated, the meadows flooded and drained in time for farmers to mow and store enough hay for winter use. Timing was crucial since flooding in the wrong season ruined winter prospects.

Finally, colonial farmers also used the Charles River watershed as a source of waterpower. In the seventeenth century, the rudimentary state of waterpower technology limited mills to the gentler tributaries of the Charles. On the Charles itself, only one solitary gristmill existed in 1650. The picture of New England's colonial water mills conveys a homely, quaint sense of charm. Setting out early in the morning, across roads of recent vintage, one can imagine farmers with oxen and carts, hauling their grain, mostly Indian corn and rye, to one of the local gristmills where it was ground into flour or meal. Sawmills with straight up-and-down saws working at one hundred strokes per minute, fulling mills with wooden mallets pounding cloth — these were simple operations that relied on small quantities of

---

[17] My debt here is to Brian Donahue. For a discussion of the Concord River meadows, see his "'Dammed at Both Ends and Cursed in the Middle': The 'Flowage' of the Concord River Meadows, 1798–1862," *Environmental Review* 13 (1989): 47–67.

[18] Howard S. Russell, *A Long, Deep Furrow: Three Centuries of Farming in New England* (Hanover, N.H., 1976), 47.

waterpower, generally not more than a few horsepower.[19] Colonial authorities commonly offered settlers inducements to build the early saw- and gristmills.[20] The town of Dedham, Massachusetts, granted Abraham Shawe, an early settler there, sixty acres of land in 1637 in exchange for constructing a corn mill.[21] In 1680, Andover, Massachusetts, in the Merrimack valley pledged free timber and land to anyone who erected a sawmill, gristmill, or fulling mill on the Shawsheen River.[22] Such early traditional mills were open to the public, in contrast to the later nineteenth-century factories that served private interests. The early mills functioned as an important part of the agricultural economy, relied on to help meet the subsistence needs for food, clothing, and shelter.[23]

Most of the work done at preindustrial mills was seasonal and did not require the continuous use of waterpower as did large nineteenth-century factories. Even so, tension surfaced during this early period over conflicting patterns of water use. Especially during dry summers, when the Charles and other New England rivers were running low, mill owners closed the floodgates on their dams to store water for power. This could send water spilling into nearby meadows at a time when farmers needed them drained to bring in their hay. But seventeenth-century mills were regulated to safeguard meadowlands. The town of Dedham protected them by requiring flashboards (wooden boards used to prevent water from flowing over the dam and thus saving it for waterpower) placed on milldams to be lowered during the summer months when drainage was necessary. Although millers were permitted to flood them in times of severe drought, even then it was "prouided still that meaddows be cared for by the Millers that Haue liberty to do this worke."[24] Whatever public benefit the mills afforded, they

[19] Hamilton, "Neponset and the Charles," 111, 116; Louis C. Hunter, *Waterpower in the Century of the Steam Engine*, A History of Industrial Power in the United States, 1780–1930, vol. 1 (Charlottesville, Va., 1979), 6. For more on the role of waterpower technology in an agricultural economy, see Hunter's discussion in chs. 1 and 2.

[20] Hunter, *Waterpower*, 28.

[21] Erastus Worthington, *Historical Sketch of Mill Creek, or Mother Brook, Dedham, Mass.* (Dedham, Mass., 1900), 1.

[22] Benno M. Forman, "Mill Sawing in Seventeenth-Century Massachusetts," *Old-Time New England: The Bulletin of the Society for the Preservation of New England Antiquities* 60 (1970): 116–17.

[23] Hunter, *Waterpower*, 50.

[24] Don Gleason Hill, ed., *The Early Records of the Town of Dedham, Massachusetts, 1672–1706*, 6 vols. (Dedham, Mass., 1886–99), 5:184.

were regulated with an eye to insuring farmers a large measure of protection for their meadow property.

By the eighteenth century, the Charles River valley inched its way toward explosive conflict. Mill owners angered both farmers and fishermen as more dams were built overflowing meadows and blocking the passage of fish. In 1738, Josiah Kingsberry, Joseph Ephraim, and others from Newton, Needham, Weston, Medfield, and Natick – including a small group of Indians – complained to the legislature about the milldam in Watertown, which they believed stopped fish from reaching them. The committee appointed to resolve the matter ruled that the stones knocked down from the dam by ice each winter should not be replaced until the first day of May to allow fish to pass upstream.[25] A balmy winter could doom the prospects for spring fishing. If, instead, the New England winter proved harsh and frigid, the mills would suffer.

Significantly, the committee enlisted nature's help in resolving this dispute. In part, such action may have simply been an attempt by the government to abdicate responsibility. Still, the outcome does suggest an attitude of deference toward the environment. The decision made both the interests of fishermen and mill owners equal before the arbitrary laws of nature. To some extent, a reluctance to tamper with the natural flow of the water is apparent here. Rather than weighing the positive utilities to be derived from using the river as either a source of fish or waterpower, the commission chose to let nature determine the course of economic development. Evidently, this was still a culture willing to concede to nature the right, at least in part, to influence how history would be made.[26]

Another source of conflict on the river emerged a decade later. In 1749, Matthew Hastings acquired a mill privilege on the Charles River in Natick and constructed a dam to supply power to a saw, fulling, and corn mill.[27] The dam apparently flooded the meadows of farmers several miles upstream in the towns of Medfield and Medway. Led by Oliver Ellis, the meadow owners had taken Hastings to court for flooding their lands and initially lost. They appealed the case to the superior court

[25] Province Resolves 1738–39, chs. 38, 39, 86, 87, 142.
[26] Arthur McEvoy suggested the ideas in this paragraph to me.
[27] Oliver N. Bacon, *A History of Natick, From its First Settlement in 1651 to the Present Time* (Boston, 1856), 156.

and won, but Hastings, who asked the court to reconsider the case, obtained a ruling in his favor. In 1753, the matter came before the provincial government of Massachusetts. Ellis and roughly eighty people from Medfield and Medway petitioned to have the superior court ruling in Hastings's favor set aside and a new trial on the flooding of the meadows granted.[28]

In their own petitions to the government, Hastings's lawyers pointed to a statute passed in 1713 that offered a "fairer method of determining controversies" over flooding.[29] That statute, referred to as a mill act, empowered a jury to assess damages to be paid each year by a mill owner for land flooded by a dam.[30] Apart from the statute, the common law itself offered a number of avenues of relief to landowners whose property was flooded. They could, for example, take the law upon themselves and abate the nuisance with their own hands. Alternatively, they could bring an action for damages against the mill owner – the recourse taken by Ellis and the other meadow owners.[31] The 1713 statute was set up to prevent actions for flooding at common law and in theory should have stopped the Medfield and Medway farmers from bringing separate actions for damages against Hastings.

Unfortunately for Hastings, the meadow owners ignored the mill act. According to Hastings's lawyers, the mill act was "despised" by Ellis and the other meadow owners who threatened to bring further actions against their client.[32] As a result, they explained, Hastings remained "confined at home only for fear of the multiplied suits of these people as there are about sixty of them who complain of him and were he to appear in this town would each take their several action[s] against him."[33] Nonetheless, Hastings did succeed in having the provincial government dismiss Ellis's petition.[34]

What does this episode suggest about the way water was conceived of and handled in the mid-eighteenth century? In sub-

[28] Province Resolves 1753–54, ch. 96; also see Petition to Governor and Council, 11 Sept. 1753, vol. 1:239–40, Mass. Archives Collection, Mass. Archive.

[29] Province Resolves 1753–54, ch. 112; see Petition to Governor and Council, 14 Sept. 1753, vol. 43:749, Mass. Archives Collection, Mass. Archive.

[30] Province Laws 1713–14, ch. 15.

[31] Morton J. Horwitz, *The Transformation of American Law, 1780–1860* (Cambridge, Mass., 1977), 48.

[32] Petition to Governor and Council, 4 Dec. 1753, vol. 1:249, Mass. Archives Collection, Mass. Archive.

[33] Petition to Governor and Council, 14 Sept. 1753, vol. 43:750.

[34] Province Resolves 1753–54, ch. 96.

stance, the 1713 mill act held out enormous hope to milldam owners. The act, had it been followed, would have freed them from the burden of costly litigation resulting from the flooding of land by their milldams. In its place, the act substituted a more orderly procedure for dispensing damage payments. Moreover, once a jury had agreed on the damages, the act prevented landowners from taking any further action on the matter. The language and implications of the act seemed to represent a shift to a dynamic vision of property more in line with the needs of a growing economy. No longer were property owners afforded absolute protection of their lands. Their property could be compromised for the sake, as the act states, "of mills serviceable for the publick good." Not surprisingly, Ellis and the other meadow owners refused to pay attention to the enactment. More important, the Massachusetts court system did not interpret the act in a way that blocked traditional common law recourse for nuisance.[35] On this point, the Massachusetts courts and the Charles River meadow owners agreed. In the mid-eighteenth century, however strong the impulse for economic development, property owners still were accorded the right to the peaceful enjoyment of their lands. Such a vision of property, it follows, acted to restrain the more aggressive, instrumental handling of water in the service of economic growth.

By the latter part of the eighteenth century, however, water was on its way to becoming a powerful tool for spurring economic development. At this time, the Massachusetts legislature showed renewed interest in encouraging the construction of mills. In 1796, the legislature passed another mill act, superseding the earlier one.[36] On the surface, both statutes seem similar: Both established a procedure for compensating landowners when a milldam was built across a stream. Both enactments also stipulated that once a jury delivered a verdict on damages for flooding, the decision would bar any further actions for damages.

But a careful reading of the 1796 act suggests some subtle changes. The substance of the 1796 mill act clearly suggests that the legislature meant it to be the *exclusive* remedy for the flooding of lands. This is evident in the far more comprehen-

[35] Horwitz, *Transformation of American Law*, 47.
[36] Act of 27 Feb. 1796, ch. 74, [1794–5] Mass. Acts & Resolves 443.

sive nature of the enactment. The 1713 statute offered a static remedy to the flooding problem in that once damages were assessed, they were simply to be paid annually.[37] The 1796 enactment, in contrast, allowed the sum of yearly damages to be either increased or decreased. In addition, unlike the earlier statute, the 1796 mill act required milldam owners to post a bond to insure payment of the damages. Finally, the 1796 act placed a cap of four pounds per year for damages resulting from flooding.[38] In sum, the later statute was far more dynamic, accounting as it did for a number of different eventualities. This feature of the legislation increased the likelihood of its functioning as an exclusive remedy for the flooding of land.

The 1796 statute was amended over the nineteenth century, overturning, in the process, the common law's protection of land from flooding.[39] According to the legal historian Morton Horwitz, the Massachusetts mill acts were construed by jurists to proscribe common law action for flooded land.[40] In general, nineteenth-century Massachusetts mill owners – unlike Hastings – could leave their homes knowing that the mill acts offered protection from the threat of litigation at common law.

The mill acts represented a more dynamic understanding of water as a form of property. By the nineteenth century, water could be used more easily than before to further economic ends. Milldam owners were now able to build dams and control water for production with limited risks of litigation for flooding. "Under the influence of the mill acts," Horwitz points out, "men had come to regard property as an instrumental value in the service of the paramount goal of promoting economic growth."[41]

While legislators in Boston were off rewriting the state's mill act, farmers and mill owners upstream on the Charles River continued to clash over milldams. During the late eighteenth century, conflict erupted along the stretch of river from Dedham downstream to Newton.

---

[37] Province Laws 1713–14, ch. 15, § 3.
[38] Act of 27 Feb. 1796, ch. 74, §§ 3, 4, [1794–5] Mass. Acts & Resolves 443–4.
[39] The 1796 statute was amended in 1798 to give mill owners the right to flood the land of others without paying damages if they proved that the complainant had no title to the land. See Act of 28 Feb. 1798, ch. 63, [1796–7] Mass. Acts & Resolves 451. Also see Act of 26 Feb. 1825, ch. 153, [1822–5] Mass. Acts & Resolves 658.
[40] Horwitz, *Transformation of American Law*, 47–51.
[41] Ibid., 53.

Arriving in Dedham from the west, the river then turns 270 degrees. The river flows slowly here. At times the water lies still, refusing to move from the meadows that flank the river at this point. Farmers had often been eager to see the water leave and attempted to goad it from their land. In 1639, the town of Dedham permitted a ditch to be dug from the Charles River to East Brook, which flowed southeast into the neighboring Neponset River. The Mother Brook diversion, as it was later called, forced water to escape from the Charles, draining the meadows of the stagnant waters that so cursed farmers.[42]

More than one farmer was often found digging in that ditch – widening it, deepening it, whatever they could do to compel better drainage. In 1766, meadow owners in Roxbury, Dedham, Needham, and Newton urged the legislature to appoint a committee to look into the flooding of two thousand acres of land lying along the Charles.[43] To resolve the problem, a sill was placed in Mother Brook to divert more water from the Charles, lowering the level of the water on the flood plain.[44]

Unfortunately, the water returned. In the spring of 1789, the Charles intruded, yet again, on the Dedham meadows. This made Aaron Fuller, Isaac Bullard, and several of their neighbors so angry that they decided to set off on horseback to confront mill owners at Newton Upper Falls. When they arrived, they tried to persuade the mill owners to remove the flashboards on their dams, which had been placed there to store water for power, lest their hay be ruined. They were unsuccessful.[45] As one observer would later conclude, it appeared that mill owners had established "a monopoly of the common highways of Nature."[46]

The lay of the land in Dedham where the Charles slows to maneuver through its series of turns made it prone to flooding. And heavy spring rains often sent water over the banks. Watersheds are sensitive, and sporadic deforestation, which increases the quantity of streamflow, could have presented an added bur-

---

42 Worthington, *Mother Brook,* 2.
43 Petition to General Court, 15 Sept. 1766, vol. 1:447, Mass. Archives Collection, Mass. Archive.
44 Worthington, *Mother Brook,* 7.
45 Nathaniel Ames Diary, 20 June 1789, Dedham Historical Society, Dedham, Mass. Ames, an owner of meadows in Dedham, lent Bullard his horse to go to Newton.
46 Eliphalet Pond, Note to the Map of the Town of Dedham Taken in the Year 1795 by Order of the General Court, 1795, Dedham Historical Society.

den.[47] But the flooding of meadows during the drier summer months was more likely the result of mills storing water for power, and farmers were at least sometimes correct in their accusations. In addition, the extensive damming of the river during the late eighteenth century aggravated the problem. More dams were constructed across the Charles between 1778 and 1794 than ever before. Five new ones were built (making a total of eight) across the lower portion of the river: one in Watertown, two in Waltham, and one each at Newton's Upper and Lower Falls[48] (see Map 1.1). Between 1775 and 1800, the number of mills depending on the Charles for power more than doubled, rising from nine to twenty. In hesitant motion, the water now paused at the traditional gristmills, sawmills, and fulling mills that normally congregated around such dams, as well as more elaborate and novel ventures – paper, dye, and snuff mills – producing for a broader market.[49]

In part, the more extensive use of the river for power stemmed from technological change. Innovations in paper-making technology, originating in the Netherlands, filtered to America by the eighteenth century. In particular, the Hollander machine brought mechanization to the washing and beating of cotton and linen rags, turning them into a liquid pulp used to make paper. Early in the nineteenth century, the perfection of the Fourdrinier papermaking machine (first shipped to America in 1827) further replaced tasks customarily done by hand.[50] It is difficult to be certain about the technology used in eighteenth-century mills. Nevertheless, four paper mills existed along the Charles by 1800, and one imagines some connection here to the changing technology.[51] Mechanization in papermaking and later in cotton textiles, particularly the Arkwright water frame, increased the demand for water. By the turn of the century, technological and economic change combined to bring more waterpowered manufacturing to the Charles River, a development that threatened the valley's farmers and fishermen.

---

[47] The effects of deforestation on streamflow are more fully discussed in Chapter 4.
[48] Thelma Fleishman, *Charles River Dams* (Charles River Watershed Association, Auburndale, Mass., 1978), 20–4.
[49] Hamilton, "Neponset and the Charles," 111.
[50] Judith A. McGaw, *Most Wonderful Machine: Mechanization and Social Change in Berkshire Paper Making, 1801–1885* (Princeton, N.J., 1987), 41, 96; on the papermaking process, also see Newton Historical Commission, *Newton's Nineteenth Century Architecture: Upper and Lower Falls* (Newton, Mass., 1982), 16–17.
[51] Hamilton, "Neponset and the Charles," 111.

Some people felt the drive to develop the river for power was shortsighted and unfair; observed one farmer, "our political Fathers will sacrifice the richest & in some cases the sole property of hundreds of good citizens" for the sake of industry.[52] At times, however, concessions were made. In 1778, David Bemis, who had recently purchased thirty-nine acres of land in Watertown, and Dr. Enos Sumner constructed a dam across the Charles and built a paper mill. When they erected the dam, Bemis opened sluices in it to allow fish to pass. But by this time the river was too obstructed to make the fishway of much value.[53]

From as far upstream as Medway right down the valley to Newton, farmers who looked forward to the spring fish runs of shad and alewives grew bitter and disappointed. A farmer living near the Wrentham Ponds in 1785 is said to have recalled that his father "used to eat fine salmon that were taken every spring at the entrance of said Ponds which communicate with Charles River but of late years their passage is stopped by mills."[54] That same year petitioners throughout the valley complained to the legislature that the "Charles River formerly afforded a passage for the salt water fish to ascend, and that such fish did annually ascend to the great benefit of the inhabitants residing near the river." But lately, they continued, "these advantages of nature are suspended, and the passage of fish, over the great falls in Newton [is] totally impeded by the dams and artificial obstructions of the water." The legislature appointed a committee to study the problem but took no action for several years.[55]

Nathaniel Ames, a Dedham physician concerned with protecting farming and fishing interests, recalled that the legislature considered a bill to secure free passage of fish up the Charles in 1790. Two members of the legislature, he explained, suggested that the passage of such an act "would spoil Eliots [sic] important Manufacture of snuff."[56] In 1778, Simon Elliot,

---

52 Note to the Plan of the Charles River, 1791, Maps and Plans, 3d series, vol. 39:11, Mass. Archive.
53 Charles A. Nelson, *Waltham, Past and Present and Its Industries* (Cambridge, Mass., 1879), 125; Resolve of 25 Apr. 1781, ch. 78, [1780–1] Mass. Acts & Resolves 413.
54 The quote is that of Nathaniel Ames, who was paraphrasing J. Whiting. See Nathaniel Ames Diary, 26–31 Oct. 1785.
55 Resolve of 1 Dec.1785, ch. 143, [1784–5] Mass. Acts & Resolves 813; Resolve of 24 Mar. 1786, ch. 202, [1784–5] Mass. Acts & Resolves 950; quotations are from Petition filed with Resolve 1785, ch. 143, Passed Resolves, Mass. Archive.
56 Nathaniel Ames Diary, Jan. 1790.

a wealthy Boston tobacco merchant, had bought five acres of land and water rights at Newton Upper Falls and over the next twelve years built four snuff mills with twenty mortars to crush tobacco.[57] By the end of the decade, Elliot's mills were reportedly one of the largest establishments of their kind in New England. He objected to the bill, and the legislature sided with him, refusing to allow it to reach a reading and sending it to a committee instead. In February 1790, Dr. Ames left Dedham prepared to address the committee, to confront them with the injustice that had spread across the Charles River valley. But when he arrived, he found the committee unwilling to rule on the matter, in his words, "denying the jurisdiction of the Gen-[eral] Court in fish matters" and raising "my indignation so that I could not express the Ideas I had conceived."[58]

That same year, farmers in Dedham and Roxbury asked for the state's help in combatting the persistent flooding of their meadows. Under a statute passed early in the eighteenth century, Massachusetts's farmers could have the state appoint a special committee, called Commissioners of Sewers, to help with the drainage of their lands. An amendment to the act passed in 1745 allowed the commissioners to open the sluices in milldams, provided they caused their owners as "little inconvenience or damage" as possible. But should such damage occur, the commissioners were to assess the proprietors of the overflowed lands for the expenses.[59] The act prejudiced the interests of the state's meadow owners, who were forced to assume the cost of draining lands whether or not milldams caused the flooding. It also betokened the day when landed property interests would not be allowed to stand in the way of economic development.

While the state appointed commissioners, neighboring farmers submitted a plan of their own. It called for the construction of five reservoirs allowing Simon Elliot to draw water for his mills at Newton Upper Falls, eliminating the need for his dam. One proponent of the idea waxed enthusiastic about its prospects, believing the removal of the dam would see "the sour sunken Meadows restored to their former sweetness & inexhaustible fertility – the fattest fish of the sea that now annually

[57] George S. Gibb, *The Saco-Lowell Shops: Textile Machinery Building in New England, 1813–1949* (Cambridge, Mass., 1950), 15.

[58] Nathaniel Ames Diary, 5 Feb. 1790.

[59] Province Laws 1745–46, ch. 16.

attempt to [ascend] the ancient ponds & sources of the river admitted to refresh remote inhabitants. . . . Agriculture & Commerce harmonizing instead of hurting each other – the cheerful citizens would bless the government that restored them to so happy a condition!"[60] But the plan failed to impress the legislature. Instead, it accepted the officially commissioned report, which held that the laws already established were sufficient to resolve the complaints of the valley's farmers. "So now we have no other recourse," explained a frustrated Ames, "but to apply to the Governor & Council for Sewers to be commissioned to order the necessary operation and assess the necessary expences" (including any occasioned by damage done to the mills in the drainage process).[61]

Near the end of the century, farmers owning meadows near the Charles formed a corporation to take over responsibility for draining their lands, lest they remain "a barren and unwholesome quagmire." As they reasoned in 1796, their incorporation was necessary, "because no one proprietor can be expected . . . to take upon himself the burden of engaging in a tedious contest with the powerful combination of millers, which, if successful, would oblige him to spend more than the value of his land."[62] By the late eighteenth century, it seemed that farmers were no longer entitled to an absolute right to freely use their land. As one person remarked in 1795, the valley's agriculturalists "find their natural rights stolen from them, and their best property at the mercy of one or two Millers, still the lucky favorites & likely to remain, so long as the rage for Factory at every place, whether others sink or swim, continues the rage of Government."[63]

In the meantime, Dr. Nathaniel Ames – the self-proclaimed voice of justice, the great defender of the public's right to the Charles – had apparently been busy rethinking his views about factories. By 1812, Ames had other concerns aside from the flooding of his meadows. Not least among them was his interest in the newly created Norfolk Cotton Manufactory. Incorporated in 1808, the company had obtained a water privilege on Mother Brook in Dedham and built a large wooden cotton mill.

---

60 Note to the Plan of the Charles River.
61 Nathaniel Ames Diary, 1791. No month and day are specified in the diary entry.
62 Act of 10 Mar. 1797, ch. 84, [1796–7] Mass. Acts & Resolves 172; quotations are from Petition, 11 Feb. 1796, filed with Act 1797, ch. 84, Passed Acts, Mass. Archive.
63 Pond, Note to the Map of the Town of Dedham.

Ames computed his share in this venture and found it to be worth $884.[64] People are perhaps forever busy redrawing the fine line of propriety – Ames no less so. Whether he realized the contradiction between his earlier stands against the mills and his later investment is impossible to say. But the conversion of such a fierce opponent of factories may suggest a broader cultural transformation that primed the way for the BMC.

Had Benjamin Harrington of Waltham decided in the fall of 1815 to wander down to the land he formerly owned along the Charles River, he would have returned home wet. The fifteen acres he once owned were now under water. So too was the land of several of Harrington's neighbors as well as property owned by people living in the nearby towns of Watertown, Newton, and Weston – close to two hundred acres in all, submerged beneath an expansive Charles River. Even after John Boies built the first dam across the Charles in Waltham to produce paper in 1738, virtually all of this land remained safe and dry. Only when Patrick Tracy Jackson, who represented the interests of the BMC, bought Boies's mill and raised the height of the dam did the water begin to spill out over the land, turning it into an enormous mill pond, perhaps the most visible sign that a new type of industrial venture had arrived on the Charles.[65]

The BMC's success at Waltham is a widely known chapter in the rise of New England textiles. Francis Cabot Lowell's industrial intrigue in Britain, his collaboration with the eminent mechanic Paul Moody, and the ultimate development of the power loom – all this is a familiar episode in the history of American business.[66] The factory's founders have attracted attention for their technical ingenuity, their financial prowess, their superb spirit of enterprise. Not as widely recognized is the subtle environmental dimension to this story of vigorous industrial expan-

---

[64] Nathaniel Ames Diary, 25 May 1812.

[65] To determine the damage to land caused by the dam, land was viewed between 8 and 15 June 1815, when water was held back by the dam, and again between 26 June and 8 July 1815, when the water was drawn off. Then the BMC purchased the affected land. See J. Gale, J. Osborne, B. Harrington, M. Wellington, S. Leaverns, J. Fuller, E. Starr, T. Beals, J. Starr, N. Fuller, Jr., A. Strong, N. Weld, J. Tucker, S. Ross, E. Leaverns, S. Bemis, S. Harrington, J. Houghton, S. Davis, J. Stedman, J. Fox, J. Robinson, Newton Selectmen, Weston Selectmen to BMC, 21 Nov. 1815, bk. 214:101–47; M. Vose and S. Sanderson to BMC, 6 Dec. 1815, bk. 214:149–52, Middlesex County Registry of Deeds, Cambridge, Mass.

[66] See, e.g., Robert Sobel, *The Entrepreneurs: Explorations Within the American Business Tradition* (New York, 1974), 1–40, or Gibb, *Saco-Lowell Shops*, 4–14.

sion. Great innovators in all respects, these entrepreneurs were no less adept when it came to redesigning the natural world. The origins of the Waltham–Lowell system, as it developed from its start on the Charles River, entailed a vital reshaping of the environment, as fundamental as the new system of labor and management that developed there; for here, the groundwork was laid for the use of water on an entirely new and different scale.

Although using waterpower for large-scale textile production seemed a novel idea in early-nineteenth-century America, across the Atlantic, Britain's rivers were already hard at work and had been for some time. Cotton textile production, fueled by the technical advances of Arkwright, Crompton, Hargreaves, and Cartwright, created an unprecedented demand for waterpower. Water is a rich and abundant resource in many parts of England and Scotland, but the enormous quest for power pressed hard against the natural capacities of the land. By roughly 1760, there were no more waterpower sites available within a five-mile radius of Birmingham. Even streams with modest falls such as the River Leen near Nottingham were well developed by the late eighteenth century.[67] Resourceful industrialists searched hard for ways to exploit the region's rivers more effectively, building dams and reservoirs to help with water control. Still, before the second decade of the nineteenth century, large reservoirs on the scale of the BMC's were rare. William Strutt's Belper mills, one of the largest groups of water-driven factories in Britain at this time, raised a dam for one mill across the Derwent that created a mill pond of only fourteen acres.[68]

Just before the Waltham venture, Francis Cabot Lowell and Nathan Appleton, the Boston merchants who would play central roles in harnessing New England's rivers for production, were both visiting Scotland. Lowell rented a house in Edinburgh while Appleton journeyed about the Southern Uplands, the two meeting at one point to discuss the prospects for textile manufacturing.[69] Although several thousand miles from New England, the rivers they saw may well have reminded them of

[67] Terry S. Reynolds, *Stronger Than a Hundred Men: A History of the Vertical Water Wheel* (Baltimore, 1983), 267–8.

[68] Ibid., 273.

[69] Nathan Appleton, *Introduction of the Power Loom, and Origin of Lowell* (Lowell, 1858), 7.

home. The Scottish landscape bore a remarkable resemblance to New England, especially in its potential for waterpower. The successive glaciations that scoured both regions created streams with frequent falls, and the climate also delivered a reasonably reliable amount of rainfall to both areas.[70] Geologic forces disposed both places to a similar industrial trajectory. Perhaps they did not plan it that way, but for these two aspiring businessmen, Scotland seemed a good choice for inspiration.

When Lowell and Appleton made their respective trips, Robert Owen's New Lanark mills were among Britain's largest water-driven mills in terms of the number of employees. Lowell may well have visited there; Appleton definitely did so.[71] On 16 October 1810, having already spent several weeks touring the country, pausing for brief spells at Glasgow, Edinburgh, and Paisley, Appleton made his way toward New Lanark. On the trip there, Appleton passed beside the River Clyde, the lifeblood of the New Lanark mills. Etching its way through the Scottish countryside, the Clyde falls in a series of spectacular cascades before it levels out upon reaching New Lanark, where the channel broadens and an island forces the river to split its water. Opposite the island were four cramped mills – two of them identical in form – perched end to end on a shelf of rock, close to the water.[72] When Appleton arrived, three of the mills were in operation. Together, they employed roughly twelve hundred people, the water driving over 24,000 spindles.[73] Gazing down at the mills, Appleton saw his future prosperity, or a small-scale version of it, opening before him. "Lanark stands on the top of a hill," wrote Appleton, "& hence we proceed to the River where are a range of very extensive cotton mills through

---

[70] Robert B. Gordon, "Hydrological Science and the Development of Waterpower for Manufacturing," *Technology and Culture* 26 (1985): 207.

[71] Sobel, *Entrepreneurs*, 20; Frances W. Gregory, *Nathan Appleton: Merchant and Entrepreneur, 1779–1861* (Charlottesville, Va., 1975), 49. Thomas Bender does not believe the Waltham–Lowell system was much influenced by the New Lanark mills. See Thomas Bender, *Toward an Urban Vision: Ideas and Institutions in Nineteenth Century America* (Baltimore, 1975), 35–6. Robert Dalzell, however, argues that what Francis Lowell saw while traveling through Scotland, a country experiencing demographic and economic transformation, must have had some impact on his later business practices at Waltham. See Robert F. Dalzell, Jr., *Enterprising Elite: The Boston Associates and the World They Made* (Cambridge, Mass., 1987), 14–25.

[72] Frank Podmore, *Robert Owen: A Biography* (1906; reprint, London, 1923), 80–1.

[73] Richard L. Hills, *Power in the Industrial Revolution* (New York, 1970), 245.

the windows of which we saw the wheels in full motion for 4 or 5 stories together."[74]

When Lowell returned home from his trip late in 1811, he sought financial support for a textile factory, securing the BMC's charter from the Massachusetts legislature two years later. Appleton was a reluctant supporter at first, contributing only half the amount asked of him.[75] The charter required the BMC to locate its factory in Suffolk County or within a fifteen-mile radius of the city of Boston. The company settled in Waltham, a town with a population of roughly 1,250 people at the time.[76] When the BMC arrived (after purchasing Boise's paper mill), Waltham hosted a small paper mill and the Waltham Cotton and Wool Factory Company, incorporated in 1812. Waltham Cotton produced textiles in a four-story wooden building, employing in 1815 about 200 people, and operating roughly two thousand spindles – a small operation soon to be overshadowed, then dominated, by its industrial neighbor upstream.[77]

By the fall of 1813, the BMC began constructing its first mill. The following January, Jackson penned Lowell a note to inform him of the progress they had made with the power loom. "I have got our loom up," wrote Jackson, "and yesterday wove several yards *by Water*."[78] Later that year, Lowell invited Nathan Appleton to Waltham to see the loom in operation. "I well recollect," Appleton recalled some forty years later, "the state of admiration and satisfaction with which we sat by the hour, watching the beautiful movement of this new and wonderful machine, destined as it evidently was, to change the character of all textile industry."[79]

---

[74] Nathan Appleton, 16 Oct. 1810, Journal of 1810, vol. 15 of bound volumes, Appleton Family Papers, MHS.

[75] Robert V. Spalding, "The Boston Mercantile Community and the Promotion of the Textile Industry in New England, 1813–1860" (Ph.D. diss., Yale University, 1963), 18.

[76] Mailloux, "Boston Manufacturing Company," 55; Howard M. Gitelman, *Workingmen of Waltham: Mobility in American Urban Industrial Development, 1850–1890* (Baltimore, 1974), 4.

[77] [Samuel Ripley], "A Topographical and Historical Description of Waltham, in the County of Middlesex, January 1, 1815," *Collections of the Massachusetts Historical Society* 2d ser., 3 (1815): 263–4.

[78] P. T. Jackson to F. C. Lowell, 30 Jan. 1814, box 7, file 14, Francis Cabot Lowell Papers, MHS.

[79] Appleton, *Introduction of the Power Loom*, 9.

Perfecting the power loom allowed the company to success-fully combine all the major processes of textile production in one factory. The Waltham mill was a paragon of industrial efficiency. Vertically integrated with cotton cards located on the ground floor, spinning frames in the middle, and power looms on top, the mill rationalized the process of textile production.[80] Although changes were made to this form, the BMC's mills remained the prototype for other large-scale water-driven textile mills throughout the nineteenth century. A pioneering venture, the BMC also assumed an aggressive stance toward the natural world that called for water on a new scale.

The BMC thirsted for waterpower at least partly to offset its most pressing need – a shortage of workers. Women were expressly recruited to work in the Waltham mills, a work force of limited physical strength relative to one made up of men. The historian David Jeremy notes that this fact inspired greater reliance on inanimate power. Not just the power loom, but the throstle used to spin yarn and the warping frame (for joining the strands of yarn so that cloth could be woven), all, according to Jeremy, required a greater dependence on waterpower.[81] Innovative in its choice of workers and technology, the company engaged in a more thorough appropriation of water, a more systematic exploitation of the natural world.

The company sought the maximum available waterpower at its Waltham site. When the Waltham Cotton and Wool Factory Company raised the height of its dam, water flowed back into the BMC's waterwheels, impeding them from turning (a problem referred to as backwater). Although it seems likely that Waltham Cotton raised the height of its dam after the BMC had done so, the BMC sued the company and won in 1815. As part of the settlement, Waltham Cotton was forced to lower its dam eighteen inches, a move that severely handicapped it. Four years later, the BMC purchased this company for $16,000. In 1821, the BMC negotiated with Seth Bemis, David Bemis's son, for the right to control the height of the flashboard on the Watertown dam, which also contributed to the backwater problem. By the second decade of the nineteenth century, the company

---

[80] David J. Jeremy, *Transatlantic Industrial Revolution: The Diffusion of Textile Technologies Between Britain and America, 1790–1830s* (Cambridge, Mass., 1981), 181.

[81] David J. Jeremy, "Innovation in American Textile Technology during the Early 19th Century," *Technology and Culture* 14 (1973): 47, 52, 76.

had consolidated its control over much of the water along the lower Charles River.[82]

Even the way the company positioned its mills betrayed a subtle, but deeply significant change in assumptions regarding water. The BMC erected its mills in Waltham, two of which were in full operation by the summer of 1819, parallel to the Charles River, precisely as Appleton had observed in New Lanark (Figure 1.1). By building in a parallel fashion, it left open the possibility for future expansion that made efficient use of the waterpower. Yet this was not generally the custom in America. Early American textile mills, such as Samuel Slater's mill at Pawtucket, were perpendicular to the river, in a manner reflecting the small-scale character of industry with limited opportunities for expansion[83] (Figure 1.2). Indeed, as the historian Barbara Tucker notes, Slater operated his factories somewhat removed from the exigencies of the market. Slater, she explains, "was no cost-conscious factory master who treated friends and laborers as commodities."[84] And his attitude toward nature, we might add, may well have reflected a less ambitious agenda. The owners of the Pawtucket and Waltham mills had different views on their endeavors, although both undoubtedly conceived of nature in instrumental terms. Still, Slater adhered to convention, choosing to look backward to the traditional small factory. What happened in Waltham was entirely different. The BMC devised a sophisticated system of production that relied on a far more thorough exploitation of waterpower.

Samuel Slater organized his mills as partnerships, shying away from the corporate form of ownership becoming more popular in the early nineteenth century.[85] The BMC, however, secured its charter from the state. Prior to this time, corporations generally had operated as quasi-public agencies. States granted corporate charters to institutions that served a public

[82] Mailloux, "Boston Manufacturing Company," 144–5; Agreement between BMC and Waltham Cotton and Wool Factory Company, Sept. 1816, vol. 1, BMC Papers, BL; Nelson, *Waltham*, 125.

[83] Gary Kulik, "A Factory System of Wood: Cultural and Technological Change in the Building of the First Cotton Mills," in *Material Culture and the Wooden Age*, ed. Brooke Hindle (Tarrytown, N.Y., 1981), 319–20.

[84] Barbara M. Tucker, *Samuel Slater and the Origins of the American Textile Industry, 1790–1860* (Ithaca, N.Y., 1984), 110.

[85] Ibid., 102.

Figure 1.1. The Boston Manufacturing Company's factories. Painting by Elijah Smith, c. 1825. The factories were built parallel to the Charles River, which is in the foreground. (Courtesy of the Lowell Historical Society.)

Figure 1.2. *View of the Old Slater Mill in 1840.* The original Slater Mill (part without the cupola), first constructed in the 1790s, is shown perpendicular to the river. (Courtesy of the Slater Mill Historic Site.)

function such as colleges and libraries. By the early nineteenth century, as the function of the corporation began to shift away from serving the common interest, it became a very attractive form of organization. Between 1800 and 1809, 15 charters were granted to Massachusetts manufacturers; in the next decade the number increased to 133.[86] The changing nature of the corporation and its growing appeal during this period played a vital role in American economic development. This occurred as part of a much broader transformation of the corporation from a public agency to a private one operating in the interests of individual gain – a shift completed during the nineteenth century.[87]

At the founding of the BMC, however, ambiguity surrounded the corporate form of ownership. Caught between a world in which corporations served public functions and one in which they operated on a more self-interested agenda, it was unclear what to expect. Many questions remained unanswered.

[86] Oscar Handlin and Mary Flug Handlin, *Commonwealth: A Study of the Role of Government in the American Economy* (1947; rev. ed., Cambridge, Mass., 1969), 162.
[87] Horwitz, *Transformation of American Law,* 112.

How much power and authority were corporations to have over their use of the environment? How would corporate ownership and control over land and water mesh with the customs and needs of New England's farmers and small mill owners? How would nature fare when confronted with an institution in which ownership did not necessarily imply the right to manage and control everyday affairs? In 1813, as the ink dried on the BMC's charter, such questions were obscured by the novelty of it all, by the prospect of a water-driven textile factory that some imagined would work for the general good of the community. But as New Englanders quested for further economic growth, the potential for conflict inherent in the corporate form became more apparent.

The Charles River had catered for some time to the needs of economic development. But as the nineteenth century wore on, this became the principal, overriding obsession. The point is evident in the systematic attempts by industry to establish exclusive rights to the streamflow. Even early in the century, the Charles had become, at points, the scene of tremendous industrial congestion. Gone was the cozy picture of water mills snuggled in among the New England forest. A far more discordant reality intruded upon the riverscape at the upper dam in Newton Lower Falls where six paper mills, two sawmills, and one fulling mill vied with one another for a share of the Charles's flow. Competition over the water necessitated formal agreements explicitly indicating who was entitled to the water. At Newton, the six paper mills had first right to the flow according to a contract signed in 1816. If the weather turned dry and the Charles ran low, the mill owners developed a system of alternating use.[88] Such agreements were intended to address the sticky problem of ownership that the Charles River, as a common resource, presented. The contracts fixed definitive rights over the streamflow. In a sense, they did for the river what the fence did for the land.

Still, agreement was not always a simple matter. When Dedham's town fathers decided to dig the ditch that became Mother Brook, they could not have foreseen all the trouble it would later cause. By the late eighteenth century, mill owners on the Charles faced their adversaries on Mother Brook in what would

---

[88] S. Elliot, S. Curtis, M. Grant, W. Hurd, C. Bemis, and J. Ware, 26 Dec. 1816, bk. 220:149, Middlesex County Registry of Deeds.

turn out to be a protracted battle. In an effort to pool their resources, both sides petitioned the legislature for the right to incorporate. The Charles River mill owners cried out over the deviation of the flow "from its natural course," especially since the river as "public property" entitled it to the protection of the commonwealth. They alleged, in particular, that the diversion had been expanded during 1767 and 1792 in total disregard of their interests.[89] In 1798, mill owners on Mother Brook and the neighboring Neponset River, which benefited as well from the diversion of the Charles, also petitioned the legislature for incorporation. General Simon Elliot, who had taken over his father's snuff mills in 1793, and Jonathan Bixby, writing on behalf of the Charles River mill owners, strenuously objected to the request of the mill owners on Mother Brook. Mother Brook, they pointed out, "has assumed an importance to which it is not by nature entitled." Moreover, should the incorporation take place, they would "have two enemies to combat instead of one," meaning both the valley's meadow owners and the mills on Mother Brook.[90] Nevertheless, the legislature approved both incorporations.[91]

There followed several decades of industrial development and legal turmoil. In 1807, Elliot wrote a letter to the town of Newton claiming that the Mother Brook diversion had "the proprietors of mills on Charles river . . . greatly alarmed for the safety of their property."[92] The town later selected a committee to aid them in their cause. Mill owners on both streams then petitioned the Massachusetts Supreme Judicial Court in 1809 to appoint officials to allocate to each their rightful share of the streamflow. Unable to determine what alterations had been made to the river, or whether they were done "by design or accident," the officials' report recommended that three-quarters of the water remain in the Charles, the brook receiving the balance.[93] Those who owned mills on the brook – disap-

---

89 Act of 12 Feb. 1798, ch. 45, [1796–7] Mass. Acts & Resolves 405; quotations are from Petition of Proprietors on Charles River to the Mass. General Court, 20 Jan. 1796, filed with Act 1798, ch. 45, Passed Acts, Mass. Archive.

90 Act of 3 Mar. 1798, ch. 77, [1796–7] Mass. Acts & Resolves 474; quotations are from Petition of Proprietors on Charles River, n.d., filed with Act 1798, ch. 77, Passed Acts, Mass. Archive.

91 Act of 12 Feb. 1798, ch. 45, [1796–7] Mass. Acts & Resolves 405; Act of 3 Mar. 1798, ch. 77, [1796–7] Mass. Acts & Resolves 474.

92 Quoted in Francis Jackson, *A History of the Early Settlement of Newton, County of Middlesex, Massachusetts, From 1639 to 1800* (Boston, 1854), 109.

93 Report of E. Brigham, J. Kendall, and L. Baldwin to Mass. SJC, 30 Sept. 1813,

pointed by the settlement – succeeded in 1826 in inducing the court not to adopt the referees' report.[94]

Meanwhile, in the intervening period, industry continued to settle in along the banks of both waterways. By roughly the time the BMC finished building its first factory on the Charles, five dams spanned Mother Brook, and factories producing cotton, paper, and nails made use of the waterpower.[95] Eventually, both sides recognized the need for some type of compromise, and each appointed a committee to negotiate a settlement. In 1831, they signed an agreement fixing the streamflow at two-thirds for the Charles, the difference diverted to Mother Brook, a solution more favorable to the mill owners on the brook.[96] That settlement brought peace to the valley.

The summer before the agreement was signed, the BMC's agent, John A. Lowell, assessed the waterpower at Waltham. He was not terribly encouraged. "The uncertainty of the water power on the Charles River," he wrote the company's directors, "is as you know a serious evil."[97] The river had evidently disappointed Lowell and his colleagues; indeed, it had disappointed for over ten years, and that despite the company's enormous prosperity. From 1817 to 1826, it paid a substantial average annual dividend of close to 19 percent.[98] Although its prospects had dimmed somewhat by 1831, the company, as of that date, still had over 8,000 spindles and more than 200 looms producing close to 2 million yards of cloth a year.[99] Still, Nathan Appleton and Patrick Tracy Jackson – the central figures behind the enterprise after Lowell's death in 1817 – began in the early 1820s to turn their interests elsewhere. The promoters of the Waltham mills had grand plans for producing textiles along more powerful waters. Their ambitions had outgrown the Charles's modest waterpower potential, a fact perhaps made worse by the fierce legal battle waged over Mother Brook.[100]

Proprietors of Mills on Charles River v. *Proprietors of Mills on Mill Creek and Neponset River*, 7 Pick. 206 (Mass., 1828), filed with court papers of the Mass. SJC, Suffolk County Courthouse, Boston, Mass.

94 Worthington, *Mother Brook*, 8–9.

95 Ibid., 7.

96 Proprietors of Mills on Mill Creek and Proprietors of Mills on Charles River, 30 Jan. 1832, bk. 96:157, Norfolk County Registry of Deeds, Dedham, Mass.

97 J. A. Lowell to Directors, 13 Aug. 1831, vol. 3, BMC Papers.

98 Jeremy, *Transatlantic Industrial Revolution*, 101.

99 State of Manufacture at Waltham, 1831, vol. 6, BMC Papers.

100 Little direct evidence exists on how the BMC reacted to the Mother Brook mat-

As the attention of Appleton, Jackson, and their associates wandered away from the valley, they left behind their mark on the Charles's waterscape. In some respects, what the BMC did to the river was without precedent. Prior to its arrival, there had been no massive efforts to control water, no 200-acre mill ponds, nothing even close to such a scale. Yet the BMC's actions were not completely unprecedented. During the late eighteenth century, in particular, the Charles River underwent substantial economic development resulting in more dams, more mills, and more production. The state legislature played an active role in this process, encouraging the economic transformation of the region's water resources. Both the passage of a comprehensive mill act and the handling of conflict along the Charles betrayed the state government's support for water-driven factories.

The drive to promote industrial growth had, in the process, reduced water to an instrument in the service of economic change. Water by the early years of the nineteenth century was well on its way to becoming a simple utility, an abstraction employed to suit economic ambitions. This transformation was a long time in the making, but by the nineteenth century at least, there were clear signs of water's emerging status as a commodity.

ter. However, in a letter to Loammi Baldwin, Jr., a well-known New England civil engineer (see Chapter 3), Patrick Tracy Jackson showed concern over the conflict. P. T. Jackson to L. Baldwin, 13 Oct. 1825, box 16, file 5, Baldwin Collection, BL.

2

CONTROL OF WATER

It is perhaps the most celebrated river valley in America's early industrial history. The Merrimack River valley encompasses 5,000 square miles of land and water. It looks elliptical on a map and measures 134 miles from north to south and 68 miles from east to west (Map 2.1). The Merrimack River itself is 116 river miles long. At Lawrence, Massachusetts, on the lower part of the river, the average daily flow reaches over 6,000 cubic feet per second – a figure roughly eighteen times that offered by the Charles River at Waltham. There is a great deal of water here, 117,382 acres of surface water present in the valley, including Lake Winnipesaukee, New England's second largest lake.[1] The valley offers a prodigious, promising waterscape, so attractive indeed that in 1821, the Boston Associates arrived here and prepared for business. Over the next quarter-decade they brought the valley's waters under their control and spurred the region on its rise to technological dominance.

The Merrimack River begins in Franklin, New Hampshire – at the confluence of the Pemigewasset and Winnipesaukee rivers – and flows south into Massachusetts and on to the Atlantic Ocean at Newburyport. Over its entire course, the river falls a modest average of 2.6 feet per mile. Yet the better part of the descent is concentrated in short distances – at six sites in particular – three of which became major textile cities.[2] The

---

[1] Figures are from Gary V. Turner et al., "Merrimack River Basin Overview" (New England River Basins Commission, Boston, 1978), 3, 13, 18. The Charles River had a mean discharge at Waltham of 368 cfs during the period from 1931 through 1969. See U.S. Army Corps of Engineers, New England Division, "Charles River, Massachusetts: Main Report & Attachments" (Waltham, Mass., 1972), appendix D-11.

[2] U.S. Army Corps of Engineers, "Merrimack Wastewater Management: Key to a Clean River" (1974), appendix 1:23; George F. Swain, "Water-Power of Eastern New England," in U.S. Department of Interior, Census Office, *Reports on the Water-Power of the United States*, vols. 16 and 17 (Washington, D.C., 1885–7), 16:24 (hereafter cited as Swain, "Water-Power of Eastern New England").

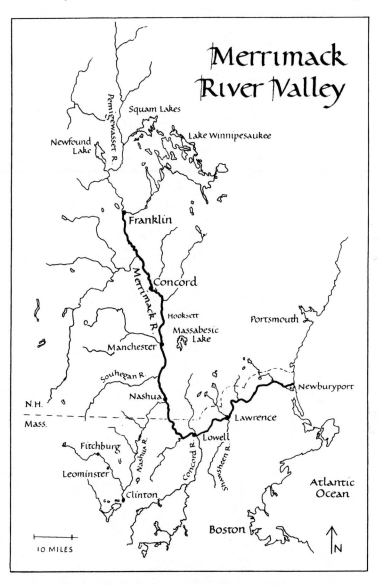

Map 2.1. The Merrimack River valley. The city of Lawrence origi-
nated at Bodwell's Falls, Lowell at Pawtucket Falls, Manchester at
Amoskeag Falls, and Concord, N.H., at Sewall's Falls. (Drawing by
David R. Forman.)

Merrimack valley presently averages forty-three inches of precipitation annually, and about half this amount ultimately
reaches the watershed's river network. The rest of the water
disappears back to the atmosphere through evaporation and
transpiration. In the northeastern corner of the valley, Lake
Winnipesaukee, Lake Wentworth, Paugus Bay, and Lake Winnisquam, each the product of glacial action occurring thousands of years ago, spread out over the land.[3] This vast supply
of water also finds its way downstream to the Merrimack River.
In all, the lay of the land and abundant waters combine to make
the Merrimack valley a rich source of waterpower. The Boston
Associates had made a wise decision to settle here.

In the fall of 1821, Nathan Appleton and Patrick Tracy Jackson became interested in a piece of land in the valley town of
East Chelmsford, Massachusetts. At the time, there was little
that distinguished this land from the nearby countryside. First
settled in the middle of the seventeenth century, the land, a
hodgepodge of woodlots, fields, orchards, scattered houses,
and a dirt road or two, was vintage New England – the kind of
place to which our thoughts turn when we think of our agricultural past. Water set the land off, completely surrounding it
like a moat rimming a castle. The Merrimack River washed
against its northern shores, while the Pawtucket Canal – built in
the 1790s to offer safe passage around Pawtucket Falls – established a rounded southern perimeter. In the early nineteenth
century, as farmers tramped back and forth across this land,
they were unmindful of the importance later generations
would attach to this spot, of the legend that would eventually
emerge here.[4]

The transformation of this rural precinct of Chelmsford into
the city of Lowell marked the start of extensive industrial development in the Merrimack valley. In little more than twenty-five
years, four major factory towns appeared in the valley, three of
them directly dependent on the Merrimack River for waterpower. By midcentury, the valley was the heartland of the large,
integrated cotton mill. At Lowell and Lawrence, Massachusetts,
Manchester, New Hampshire, and to a more limited extent at
Nashua, New Hampshire, the built environment changed radically as mills, boardinghouses, canals, and streets set off these
areas from the predominantly rural landscape that stretched

[3] Turner et al., "Merrimack River Basin Overview," 13–14.
[4] J. G. Hales, "A Plan of Sundry Farms & c. at Patucket in the Town of Chelmsford,"
(1821) in *A Portfolio of Historical Maps of Lowell 1821–1914* (Lowell, n.d.).

across New England. In 1820, before the Boston Associates descended on Chelmsford, roughly 1,500 people resided in the town. Thirty years later, Lowell's population had increased to over 33,000, making it the second largest city in Massachusetts. Likewise, Manchester's population rose from 761 in 1820 to close to 14,000 by the middle of the century.[5]

The valley's four major textile towns were largely the work of the Boston Associates, a group of men who began as merchants and later turned toward investments in the New England textile industry. Appleton, Jackson, and Francis Cabot Lowell were well-known members of the group, which by 1845 totaled roughly eighty people. The Boston Associates was not a formal organization but a network of individuals and families joined by bonds of marriage, friendship, and, of course, finance. Textiles were the focus of this complicated financial group, but the people involved did not limit themselves to this one area of investment. They were also prominent in banking, insurance, philanthropy, and politics. By the middle of the century, they had interests in thirty-one textile concerns, controlling about one-fifth of the cotton spindles operating in America.[6] New England's rich supply of water met their manufacturing needs, their factories settling along the Androscoggin, Saco, Salmon Falls, Cocheco, Chicopee, Taunton, Connecticut, and Merrimack rivers. Most of these men were not involved in the daily operation of the textile mills, but the agents and engineers they employed altered New England's waterscape, especially the Merrimack's, to meet the demands of production.

The rise of the Waltham–Lowell system in the Merrimack valley involved a fundamental transformation in the scale, degree, and purpose of water control. This meant a change in behavior, a shift in the way the Associates and those whom they employed made use of the environment. At the same time, however, the Associates' attitudes toward nature were evolving as well. How, in short, did these men perceive the natural world? A parallel process of change in attitudes and behavior happened during this period. The Associates were exercising more command over the Merrimack's water and also perceiving nature in a way consistent with that behavior.

* * *

[5] U.S. Census of Population, 1820, 1850.
[6] Robert F. Dalzell, Jr., *Enterprising Elite: The Boston Associates and the World They Made* (Cambridge, Mass., 1987), x, 79.

At the time that the Boston Associates founded Lowell, agriculture still dominated the New England landscape. But beneath this landscape lay a variety of self-generated problems that imperiled, at least in some areas, this way of life. Especially in the older, coastal areas of settlement, overpopulation and a tradition of partible inheritance threatened to fragment landholdings to where agriculture no longer offered economic opportunity for the rising generation. Population pressure, inheritance customs, and soil exhaustion caused by overfarming together may have upset the stability of agriculture by the middle of the 1700s. New Englanders dealt with this challenge to their livelihood in a number of ways. One strategy adopted by some patriarchs entailed giving the majority of the land to one or two sons to help preserve the economic viability of the property. The balance of the land generally was divided among the remaining sons who were also given assistance in either migrating afar and purchasing land, setting up a trade, or getting an education.[7]

Since the Merrimack valley spreads across a wide section of New England, some of the valley towns, not surprisingly, experienced just such pressures. In at least two Massachusetts towns, Concord and Andover, an imbalance developed between population and available resources, beginning in the 1720s in Concord.[8] But even so, the Merrimack valley did not experience a general crisis of subsistence. Not until the third quarter of the eighteenth century did the central part of the valley in New Hampshire receive an intensive flow of population.[9] Population densities in this area probably remained below those of the more thickly settled parts of the valley in eastern Massachusetts throughout the early nineteenth century.[10] Further, whatever

---

7   For a discussion of the ecological tensions that developed in New England agriculture in the eighteenth century, see Carolyn Merchant, *Ecological Revolutions: Nature, Gender, and Science in New England* (Chapel Hill, N.C., 1989), 185–90. Also see Kenneth A. Lockridge, "Land, Population and the Evolution of New England Society, 1630–1790," *Past and Present* 39 (1968): 62–80, and Philip J. Greven, Jr., *Four Generations: Population, Land, and Family in Colonial Andover, Massachusetts* (Ithaca, N.Y., 1970), ch. 8.

8   Greven, *Four Generations*, ch. 8; Robert A. Gross, "Culture and Cultivation: Agriculture and Society in Thoreau's Concord," *Journal of American History* 69 (1982): 44; idem, *The Minutemen and Their World* (New York, 1976), ch. 4.

9   D. W. Meinig, *Atlantic America, 1492–1800*, The Shaping of America: A Geographic Perspective on 500 Years of History, vol. 1 (New Haven, 1986), 289–90.

10   Norman W. Smith, however, believes that at least parts of New Hampshire were well settled by the late eighteenth century. See "A Mature Frontier: The New Hampshire Economy, 1790–1850," *Historical New Hampshire* 24 (1969): 3.

the demographic and environmental pressures, New Englanders were resourceful in their responses, turning to other means
of sustaining themselves.[11]

Nathan Appleton was the product of this set of historical
forces. The Appleton family had its roots in Ipswich, Massachusetts, where the imbalance between population and land
was at issue when Appleton's grandfather came of age in the
first half of the eighteenth century. Appleton's father, one of
eight children facing an even more arduous set of circumstances, moved to the Merrimack valley town of New Ipswich,
New Hampshire, in 1750. The historian Robert Dalzell explains
that while it took four generations to upset the equilibrium
between population and resources in Ipswich, the same development occurred much more rapidly in New Ipswich. There
were several reasons for this, but one is especially noteworthy:
The market for agricultural produce appears to have engaged
more farmers by the second half of the eighteenth century. In
turn, this probably put increasing pressure on the market for
land, as farmers produced more exclusively for sale.[12] As Appleton and his brothers came of age, one son was left to carry
on at the New Ipswich homestead. The others searched elsewhere, either taking up farming in less settled parts of New
England or turning to the world of commerce. The latter was
Appleton's path. He followed his brother Samuel into the business world, working at first as a bookkeeper before forming a
partnership with him in a mercantile firm.[13]

When Appleton returned to the Merrimack valley in the
1820s, it was, of course, not to farm the land. Perhaps the same
forces that drove Appleton into commerce and manufacturing
in some way prepared the valley for the changes engineered in
large part by him. The question of a labor force is central here.
The records of one Lowell textile company show that in 1836
large numbers of workers hailed from New Hampshire towns –
many directly from the Merrimack valley – experiencing de-

[11] See the discussion of this issue in Fred Anderson, "A People's Army: Provincial
Military Service in Massachusetts during the Seven Years' War," *William and Mary
Quarterly* 3d ser., 40 (1983): 518.

[12] Dalzell, *Enterprising Elite*, 116–19. Dalzell also notes that towns settled after the
seventeenth century matured more quickly as promoters sold off land for profit. For
an important discussion of New England agriculture, see Bettye Hobbs Pruitt, "Self-
Sufficiency and the Agricultural Economy of Eighteenth-Century Massachusetts,"
William and Mary Quarterly 3d ser., 41 (1984): 333–64.

[13] Dalzell, *Enterprising Elite*, 120. Frances W. Gregory, *Nathan Appleton: Merchant and
Entrepreneur, 1779–1861* (Charlottesville, Va., 1975), 9, 17.

mographic strain. A close link existed between the emerging factory towns and the surrounding countryside, which, until the 1840s, provided the majority of the labor force. The rural–urban migration that occurred had its roots in the changing agrarian economy. Some New Englanders, buffeted by environmental pressures stemming from a growing population, may have experienced industrial change with relief.[14]

But more than available land concerned those who settled the central Merrimack valley late in the eighteenth century and after. Since much of New Hampshire remained economically undeveloped and isolated from urban markets such as Boston and Portsmouth, attention focused on the question of transportation. And here, water figured prominently. The portrayal of New Englanders as rugged individualists, farming the land in complete self-sufficiency, appears to be largely a myth, especially by the eighteenth century. Instead, interdependent ties joined farmers to one another and to regional markets.[15] This fact lay behind the attempt to make the Merrimack River a navigable waterway beginning in the late eighteenth century.

When the Boston Associates arrived in the 1820s, they did not find untouched, free-flowing water here in the Merrimack valley. The period between 1792 and 1814 witnessed a sustained attempt to control the waters of the Merrimack River for the purposes of navigation. In 1792, at the instigation of several Newburyport merchants, the Proprietors of Locks and Canals on Merrimack River (PLC) was chartered to construct a transportation canal around Pawtucket Falls to aid in the shipping of lumber and produce downstream.[16] Before the canal, rafts of lumber were broken up at the head of the falls, hauled with oxen around it, and placed back in the water.[17] The Pawtucket Canal, when completed in 1796, etched a wide arc in the land to avoid the rocky terrain on the south side of the falls. Measuring almost eight thousand feet in length, the canal diverted water from the Merrimack into the Concord River, allowing rafts of

[14] Thomas Dublin, *Women at Work: The Transformation of Work and Community in Lowell, Massachusetts, 1826–1860* (New York, 1979), 27–30. Dublin explores the rural-urban network in "Rural-Urban Migrants in Industrial New England: The Case of Lynn, Massachusetts, in the Mid-Nineteenth Century," *Journal of American History* 73 (1986): 623–44.

[15] Pruitt, "Self-Sufficiency and the Agricultural Economy," esp. 351–4.

[16] George S. Gibb, *The Saco-Lowell Shops: Textile Machinery Building in New England, 1813–1949* (Cambridge, Mass., 1950), 65.

[17] J. Francis, Pawtucket Canal, 1876, vol. DB-8:392–4, PLC Papers, BL.

lumber to bypass the falls (the canal was not generally used to ascend the river).[18]

The Pawtucket Canal proved financially unsuccessful largely because of the construction of a competing canal. The Middlesex Canal, linking the Merrimack at Middlesex Village with Charlestown, Massachusetts, twenty-seven miles away, was finished in 1803. Establishing a transportation connection between Boston and the Merrimack valley, the new canal siphoned off much of the trade that otherwise would have reached the seaport town of Newburyport via the Pawtucket Canal.[19]

The completion of the Middlesex Canal in 1803 marked the start of a drive to make the river's many falls passable. By 1814, a system of locks and canals allowed river traffic to pass freely between Concord, New Hampshire, and Boston.[20] The opening of the river to navigation entailed a great deal of obstruction and diversion of water to circumvent the many waterfalls. In short, the needs of navigation compelled a substantial degree of water control.

This is the path the water traveled in 1815 from the river's start in Franklin, New Hampshire:[21] The first twenty-six miles were free and clear. But after winding its way in a wide arc around the town of Concord, New Hampshire, the water was detained by a dam that sent it into a series of channels one-half mile long and washed out Turkey Falls. One mile below, a dam 450 feet in length and ranging between 7 and 12 feet in height carried the water through a canal roughly one-third of a mile in length. The river then flowed freely for six miles. At Hooksett Falls, the water was diverted by another dam – constructed from a large ledge of rocks to an island in the middle of the river – into a short canal. Now eight more miles of uninterrupted travel ensued. Next, a series of dams spanning large rocks at Amoskeag Falls led the water through a one-mile-long canal. Having passed through the Amoskeag Canal, the water

---

18 Ibid.
19 The PLC was in financial trouble as early as 1803, just as the Middlesex Canal was nearing completion. See Directors' Records, 2 Dec. 1803, vol. 1, PLC Papers.
20 Christopher Roberts, *The Middlesex Canal: 1793–1860* (Cambridge, Mass., 1938), 124–35.
21 This account is drawn from a pamphlet entitled *A Report to the Directors of the Middlesex Canal, October 14, 1815*, which is reprinted in Timothy Dwight, *Travels in New England and New York*, ed. Barbara Miller Solomon, 4 vols. (Cambridge, Mass., 1969), 1:294–6; Roberts, *Middlesex Canal*, 130.

Figure 2.1. Rafts on the Merrimack River. Lithograph, c. 1850. (Courtesy of the New Hampshire Historical Society.)

was channeled around six falls in the next nine miles. The water then descended for five miles unhindered. A dam and short canal at Cromwell's Falls deflected the water from its natural path before passing it back into the river for fifteen miles of uninterrupted travel. Again, at Wicasee Falls the water snaked through another canal. After leaving the Wicasee Canal, the water passed down the Merrimack toward the Pawtucket Canal.

This elaborate scheme for diverting and obstructing the water allowed boats to navigate the river between the spring and early winter when ice closed the route. Built out of both oak and pine, the boats were typically flat-bottomed, seventy-five feet long, nine feet wide, and capable of carrying twenty tons of cargo. Salt, cement, leather, glass, lime, and iron were common products shipped upstream, while lumber dominated the downstream route[22] (Figure 2.1). The canaling of the river also benefited passenger travel, making possible, we might add, Thoreau's 1839 journey.[23]

Before the Boston Associates arrived in the valley, the river's water had been controlled to a large degree. What the Associates did with the water furthered this same process of re-engineering the natural world. Their waterpower infrastructures were built on a vaster scale, improved the degree of water control, and most important, were designed with an entirely different aim in mind. Not commerce or navigation, but production was at the heart of their efforts at harnessing water for energy.

In 1821, Appleton and Jackson began considering the Merrimack valley as a suitable place for using water to produce textiles. Buoyed by their success at Waltham, they rode through northern New England eager to expand, searching for waterpower to manufacture printed calicos. In the middle of September, on advice from Charles Atherton, they visited a spot on the Souhegan River six miles from where it meets the Merrimack River. Atherton, a lawyer living in Amherst, New Hampshire, offered to sell them his mill privilege for one thousand dollars. When they arrived, they saw the Souhegan approach the Merrimack from the west and drop twenty-eight

[22] George Stark, "Frederick G. Stark and the Merrimack River Canals," *Massachusetts Magazine* 2 (1886): 400–401.
[23] Roberts, *Middlesex Canal*, 125–7.

feet. They explored the wooded banks that enveloped the site, but found the waterpower, in Appleton's words, not "sufficiently large for our purposes."[24] When they left the valley, they consigned this place to obscurity, to what became a mere footnote in New England's industrial history.[25]

Early the following month, Jackson paid Appleton a visit at his counting room in Boston. They discussed the spot of land in the town of Chelmsford, wedged between the Merrimack River and the Pawtucket Canal. Jackson explained that Thomas M. Clark, a Newburyport merchant on the board of directors of the PLC, assured him the land and the needed stock in the PLC – devastated financially by the success of the competing Middlesex Canal – were available for purchase.[26] With the consent of Appleton and Jackson, Clark purchased land in November 1821 from a number of East Chelmsford farmers. By the end of the month, Clark controlled over 350 acres, which Jackson, Appleton and their colleagues soon developed into Lowell, Massachusetts.[27]

As the first attempt to harness the waters of the Merrimack for large-scale textile production, Lowell drew acclaim both at home and abroad. Harriet Martineau, Charles Dickens, Henry Clay, Andrew Jackson, and James K. Polk all visited this center of industrial activity. Many visitors expressed awe at the sheer productive capacity of the place.[28] Dickens noted how "the very river that moves the machinery in the mills . . . seems to acquire a new character from the fresh buildings of bright red brick and painted wood among which it takes its course."[29] Others marveled at the "clock-work movements" of the large, inte-

---

24 "Autobiography of Nathan Appleton," n.d., box 13, file 10, p. 16; C. H. Atherton to N. Appleton, 20 Sept. 1821, box 3, file 10; N. Appleton Deposition, 15 Oct. 1852, box 7, file 10, Appleton Family Papers, MHS.

25 In retrospect, it was a good decision to reject the site. The Souhegan River compared unfavorably with the Merrimack in terms of volume of flow. In addition, the Souhegan had no large lakes upstream, as did the Merrimack, for providing a reliable year-round flow of water.

26 N. Appleton to J. A. Lowell, 23 May 1848, box 7, file 2, Appleton Family Papers; Nathan Appleton, *Introduction of the Power Loom, and Origin of Lowell* (Lowell, 1858), 17–19.

27 See the following deeds: N. Tyler to T. Clark, 2 Nov. 1821, J. Fletcher to Clark, 21 Nov. 1821, M. Cheever to Clark, 21 Nov. 1821, J. Fletcher to Clark, 21 Nov. 1821, Cheever to Clark, 28 Nov. 1821, vol. 53, MMC Papers, BL.

28 See Thomas Bender, *Toward an Urban Vision: Ideas and Institutions in Nineteenth Century America* (Baltimore, 1975), 40–1.

29 Quoted in Dalzell, *Enterprising Elite*, 45.

grated cotton mills.[30] Such outward signs betrayed a deep commitment to an efficient method of production. Mass production is a distinctively American contribution to the modern world, and the Waltham–Lowell system is part of this American penchant for systematizing the workplace.[31] The physical organization of the Lowell mills reflected this. In general, the Lowell mills, and the others built on this model throughout the valley, housed waterwheels in the basement, carding machines on the first floor, spinning equipment on the second, and weaving on the top two floors. This allowed production to flow smoothly throughout the mills.[32]

Those who organized Waltham–Lowell-style mills were dedicated to standardization. For the most part, the mills produced standardized low- or medium-quality fabrics that were easy to manufacture and sell. Their most famous product was a heavy sheeting made of number fourteen yarn. Moreover, a standardized line of goods could be used to target a mass market.[33] In addition to New England, the textiles were shipped across America to markets in New York, Philadelphia, New Orleans, and on to points in Texas and the Midwest. The cotton sheetings, shirtings, and drillings eventually were marketed on a global scale, reaching Mexico, Brazil, Chile, Argentina, the East and West Indies, Turkey, Africa, and China.[34] Standardized products not only helped to take advantage of such mass markets, but also encouraged the development of machines with a uniform design. Little evidence exists before 1840 for making reliable observations on the machinery used in the mills. But what evidence is available suggests that standardized designs may well have existed. Three surviving twisting frames, for example, built roughly between 1830 and 1840 for Lowell's Merrimack Manufacturing Company, are identical.[35]

---

[30] See Herbert G. Gutman, *Work, Culture & Society in Industrializing America: Essays in American Working-Class and Social History* (New York, 1976), 26.

[31] John B. Rae, "The Rationalization of Production," in *Technology and Western Civilization*, ed. Melvin Kranzberg and Carroll W. Pursell (New York, 1967), 37.

[32] David J. Jeremy, "Innovation in American Textile Technology during the Early 19th Century," *Technology and Culture* 14 (1973): 51–2, n. 20. According to Jeremy, there were exceptions to this arrangement among the Waltham–Lowell-style mills. He also notes that the principle of flow production was three hundred years old by the early nineteenth century.

[33] Ibid., 49–50; Appleton, *Introduction of the Power Loom*, 11–12.

[34] Gregory, *Nathan Appleton*, 225–7.

[35] Jeremy, "Innovation in American Textile Technology," 51, n. 17.

The urge to set standards transcended output and machinery. Lowell was designed to encourage a well-ordered sense of existence among those who worked there, to fight off the moral degradation that many feared would accompany industrial transformation. This was accomplished through strict regulation of the work force. Workers at Lowell worked an average of twelve hours per day, six days a week, and aside from Sundays were allowed only three holidays. Management forbade drinking on company property and at least some companies required public worship. Before the massive influx of immigrants to Lowell after 1845, residence in the company-built boardinghouses was, with few exceptions, mandatory, and the women were forced to observe a ten o'clock curfew.[36] In short, a system of discipline evolved that enforced a discrete set of standards. Regularity, order, and control – these were the prime concerns of those who managed Lowell and the other large-scale textile mills in the Merrimack valley.

The built environment expressed more subtly a commitment to these values. In general, the textile mills constructed in the early nineteenth century were simple affairs. The early mills at Lowell and Nashua were roughly 150 feet long, 45 feet wide and anywhere from four to six stories high.[37] (Mills built later in the century were often larger. In 1844, the Nashua Manufacturing Company erected a mill just short of 200 feet long and 50 feet wide, while three years later the Amoskeag Manufacturing Company in Manchester constructed a building 260 feet in length.)[38] All the mills were made of red brick and were generally set off into mill yards with a bell tower often capping the central building in the complex. The design and placement of the mills and boardinghouses betrayed more than a narrow functionalism. Rather, the mills suggest that those responsible for them, deliberately or not, exerted a strong desire to carefully control the built environment. Ithamar A. Beard, paymaster of one Lowell company, may not have been instrumental in designing the city, but he did capture the prevailing sentiment. In giving his thoughts on the appropriate organiza-

---

[36] Dublin, *Women at Work*, 59, 78.

[37] See the figures in John Hayward, *The New England Gazetteer Containing Descriptions of All the States, Counties and Towns in New England*, 2d ed. (Concord, N.H., 1839), unpaginated, see under Lowell, Mass., and Nashua, N.H.

[38] Charles J. Fox, *History of the Old Township of Dunstable* (Nashua, N.H., 1846), 204; C. E. Potter, *The History of Manchester, Formerly Derryfield, in New-Hampshire* (Manchester, N.H., 1856), 568.

tion of society in factory towns, he stated that "the first thing in every society is order; without it nothing can be done."[39] The urban space created at Lowell and throughout the Merrimack valley seemed to adhere to just such a principle.

The Boston Associates manufactured more than textiles; they produced an entirely new working and living environment for those who came to Lowell. More important for our concerns here, they transformed the natural environment as well. In the fifteen years before Thoreau's trip on the Merrimack, a new waterscape emerged in Chelmsford. Having gained control of the PLC and the Pawtucket Canal, the Boston Associates created a new company. Founded in 1822, the Merrimack Manufacturing Company was formed to manufacture and print cotton textiles. The company took possession of the land and water rights near Pawtucket Falls and made plans for constructing power canals and textile factories. As these plans progressed, the natural environment at Chelmsford, soon to be Lowell, was radically altered. The waters of the Merrimack River underwent a striking change as they were redesigned to conform more closely to the needs of large-scale textile manufacturing.

Three men – Patrick Tracy Jackson, Paul Moody, and Kirk Boott – were responsible for creating the initial water delivery system at Lowell (additions to it were made in the 1840s). Moody and Jackson both had been closely associated with the development and maintenance of waterpower at the Waltham mills.[40] Boott, however, was not involved with the Boston Associates until the Lowell venture. Before then, he had traveled a number of paths. He spent a period at Harvard but never got a degree, served with the British army, and briefly worked at his father's mercantile business – in all a rather undistinguished record. But at the age of thirty-two, Boott's fortune was about to change. He was approached by Jackson and asked to become the agent for the Merrimack Company.[41] Boott began his work

---

39 Quoted in Bender, *Toward an Urban Vision*, 110.

40 Before coming to the BMC, Moody was involved with the Amesbury Wool and Cotton Manufacturing Company located in the Merrimack valley. He undertook the venture with a machinist and manufacturer named Ezra Worthen, who is said to have recommended the substantial water resources at Chelmsford to him. See W. R. Bagnall, "Sketch of the Life of Ezra Worthen," *ORHA* 3 (1884): 37, 39.

41 Brad Parker, *Kirk Boott: Master Spirit of Early Lowell* (Lowell, 1985), 30–43. Boott and his brother John W. Boott each subscribed for 90 shares of stock in the Merrimack Company. Jackson and Nathan Appleton, however, each held 180 shares, owning

as agent by purchasing over one hundred acres of land in Chelmsford, which he transferred to the company.[42] He was later appointed treasurer and agent of the PLC, a post he held until his death in 1837.

What little is known about Boott suggests his penchant for rigorous standards and discipline – a man remembered for his commanding presence. A bit of folklore, although quaint, is revealing: When at one point workers were unable to coax water into a newly excavated channel, it was suggested that Boott's hat and walking stick be brought to correct the problem.[43] Boott, together with Jackson, Moody, and a crew of millwrights and surveyors, supervised construction of Lowell's waterpower labyrinth – a system that by 1836 included seven canals and a network of supporting locks and dams for governing the Merrimack's flow.[44]

A team of Irish laborers were recruited from Charlestown, Massachusetts, in 1822 to begin enlarging the Pawtucket Canal.[45] To divert water into the canal, a dam had to be built across the Merrimack above the falls. Having purchased land and water rights, the Merrimack Company already owned at least one dam jutting out from the north side of the river. In 1824, J. B. Varnum, the owner of land opposite Chelmsford in the town of Dracut, began building another dam below the one owned by the Merrimack Company. Varnum's motives are not entirely clear. But whatever his intentions, he threatened the Merrimack Company's dominion over the water. The company's stockholders ordered Boott to inform Varnum of the company's claim to "all the water above our dam at the head of the falls to be turned into our present canal or any other canals or flumes." In 1825, the company also took steps to build a temporary dam, making clear its intention to control the flow of water.[46]

---

360 of the original 600 shares of stock. See Robert V. Spalding, "The Boston Mercantile Community and the Promotion of the Textile Industry in New England, 1813–1860" (Ph.D. diss., Yale University, 1963), 35.

[42] K. Boott to MMC, 1 Mar. 1822, vol. 53, MMC Papers.

[43] Richard A. McDermott, "The Claim to Power: The Foundations of Authority in American Industry, Lowell, 1820–1850" (Ph.D. diss., Brandeis University, 1985), 114–15.

[44] Peter M. Molloy, ed., *The Lower Merrimack River Valley: An Inventory of Historic Engineering and Industrial Sites* (Washington, D.C., 1976), 69–78.

[45] Brian C. Mitchell, "Immigrants in Utopia: The Early Irish Community of Lowell, Massachusetts 1821–1861" (Ph.D. diss., University of Rochester, 1981), 6.

[46] W. Dutton to K. Boott, Directors' Records, 23 Oct. 1824, vol. 74, MMC Papers. The

The following year, all the property and water rights owned by the Merrimack Company were transferred to the PLC, which now began managing the waterpower and land at Lowell.[47] By 1830, the PLC built a permanent dam made of wooden cribs filled with stone and spanning roughly 1,000 feet across the river.[48] But even this structure proved inadequate to meet the waterpower needs of the Lowell mills. Between 1830 and 1833, the Boston Associates formed four new textile enterprises – the Middlesex, Suffolk, Tremont, and Lawrence companies – creating a surge in the need for water. To provide a more adequate supply of water, two feet of granite were added to the dam in 1833.[49] With this addition, the Pawtucket Dam averaged about 15 feet in height, deadened the river's current for eighteen miles, and created a pond of water upstream from the falls of roughly 1,100 acres.[50]

The water in this artificial pond sat thirty-one feet above the foot of Pawtucket Falls. Maximizing the productive potential of this water became the central task of the PLC. It would have been a simple job had the company been able to construct one canal lined with mills parallel to the river, taking advantage of the water as it splashed down to the foot of the falls. But this design was not to be. As pointed out, the land south of the falls is hard and rocky, the reason the Pawtucket Canal was built in a wide pass around the falls. Instead, the plan adopted by the PLC proved ingenious, if not terribly elegant. It consisted of a complex system of power canals gouged out of the Lowell landscape, a network of channels springing off the Pawtucket Canal on which the Boston Associates pinned their hopes for power and production (Map 2.2).

When Henry Thoreau made his trip in 1839, he ignored Lowell and its canal system. But had he ventured there, he would have seen one of the more powerful examples then in existence of humankind's command over nature. By the mid-1830s, the Pawtucket Dam sent water into a maze of long waterways that

letter from Dutton to Boott makes reference to two dams owned by the MMC. Also see Swain, "Water-Power of Eastern New England," 30. Swain notes that in 1821 a wing dam existed at the head of the falls, as well as a sawmill and a gristmill. In 1825, there was a temporary dam across the Merrimack.

47 This development is discussed further in Chapter 3.
48 Swain, "Water-Power of Eastern New England," 30–1.
49 Ibid., 30.
50 Ibid., 31; also see Molloy, *Lower Merrimack River Valley,* 80–1.

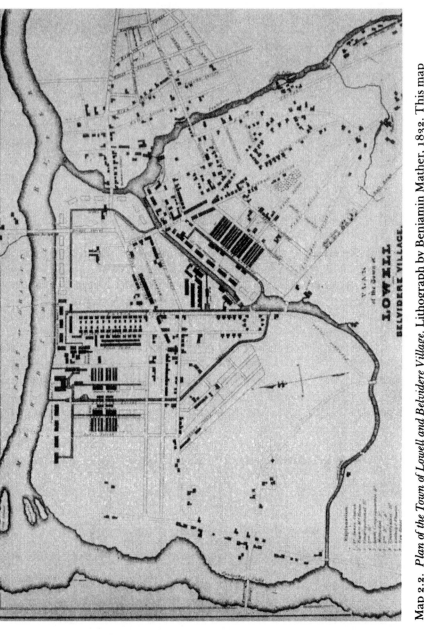

Map 2.2. *Plan of the Town of Lowell and Belvidere Village.* Lithograph by Benjamin Mather, 1832. This map shows the Pawtucket Canal, which begins on the extreme left side. At center, the canals from left to right are the Western, Merrimack, Lower Pawtucket, and Hamilton. The section marked "Contemplated Canal" at upper right eventually became the Eastern Canal. (Courtesy of the Lowell Historical Society.)

provided energy to twenty-six textile mills.[51] Before the water found its way back to the Merrimack, it performed some outstanding acrobatics.[52]

The water traveled over a mile down the Pawtucket Canal before reaching Swamp Locks where several tributary canals branched out across the land. Some of the water was conducted straight down the Merrimack Canal, plunging thirty feet through the Merrimack Company's waterwheels. (The natural fall at Lowell is actually thirty-one feet, but the water dropped one foot as it traveled from the start of the Pawtucket Canal to the Merrimack mills.) A portion of the water traveled a different path, veering sharply right into another canal, where it fell thirteen feet through the waterwheels of the Lowell Company before draining into the lower part of the Pawtucket Canal.

Some of the water charted a different course from the locks. Turning left into the 3,500-foot Western Canal, it fell 13 feet through the wheels of the Suffolk and Tremont factories. The water exiting the mills here did not fall immediately back into the Merrimack and travel out to sea. Instead, still 17 feet above the river, the water went left into another canal and turned the wheels of the Lawrence Manufacturing Company before completing the detour around the falls. Two levels of canals were involved: In one, the water fell 13 feet to work mills located at the upper level before dropping 17 feet through the mills below. Only the Merrimack Canal made use of the entire 30 feet of fall. The others operated on the two-level system.

From Swamp Locks, the water traveled two other paths. Either it made a sharp right and passed through the Hamilton Canal or a soft right down the lower part of the Pawtucket Canal. Down the Hamilton, it worked the wheels of the Appleton and Hamilton mills. Churning through the waterwheels of the Hamilton Company, the water fell into the lower Pawtucket Canal where it met water traveling this route directly from Swamp Locks. Joining forces, it journeyed to the Lower Locks where some of it parted two thousand feet down the Eastern Canal to work the wheels of the Boott mills, and the rest fell seventeen feet to the Concord River, allowing boats to

[51] Patrick M. Malone, *Canals and Industry: Engineering in Lowell, 1821–1880* (Lowell, 1983), 6.

[52] The following description of the Lowell canal system is from Malone, *Canals and Industry,* 3–8; Swain, "Water-Power of Eastern New England," 31; and Molloy, *Lower Merrimack River Valley,* 69–81.

Figure 2.2. *East View of Lowell, Mass.* Drawing by J. W. Barber; engraving by E. L. Barber, 1839, the year of Henry Thoreau's voyage on the Merrimack River. (Courtesy of the Museum of American Textile History.)

pass down the Pawtucket Canal, eventually meeting the waters of the Merrimack below the falls. Finally, the Middlesex mills, unique in getting their waterpower from both the Merrimack and Concord rivers, appeared south of where the Pawtucket Canal met the Concord. A long and arduous detour around Pawtucket Falls had ended.

Compared to prevailing attempts at controlling water in New England, the Lowell system was indeed a remarkable achievement. There were, of course, other notable efforts to control the flow of the region's waters at the time. On the nearby Concord River, Oliver Whipple, a gunpowder manufacturer, built a power canal roughly one thousand feet in length that took water from a dam at the head of Wamesit Falls.[53] On the Mill River in Massachusetts, the Springfield Armory produced arms for the federal government. In the early nineteenth century, the armory had eighteen waterwheels at three different points on the river; according to a report published in 1830, twenty-seven waterwheels using sixty cubic feet per second (cfs) of water were in operation.[54] But the Lowell mills went to greater lengths to control water, vastly overshadowing these operations. The Springfield Armory, Whipple's mills, and others like them were aggressive in the way they used rivers. Yet few mill towns, aside from those formed by the Boston Associates, were quite as thorough and matter-of-fact as Lowell in their drive to master water (Figure 2.2).

The Boston Associates succeeded at Lowell in altering nature to meet their ambitions for production. Their vision of the natural world, their understanding of the environment and how it should be used, shaped their efforts to control water. As much as anything, the vast infusion of industry into Chelmsford involved a shift in the way nature was perceived. The transition from a world where humankind remains at the mercy of nature to one where this relationship is reversed is surely one of the most powerful pillars on which industrial culture rests. In the modern world, humankind is no longer passive in its relations with nature. Instead, natural constraints are overcome by people intent on exercising their claims to autonomy and independence. Under industrial capitalism, there are many firm

---

[53] Molloy, *Lower Merrimack River Valley,* 69.
[54] Robert B. Gordon, "Hydrological Science and the Development of Waterpower for Manufacturing," *Technology and Culture* 26 (1985): 217.

believers in human ascendancy, of people set to emancipate their society from ecological dependence, from the restraints of what had once seemed a far more imposing environment.[55]

Historian Arthur Ekirch writes: "The concept that man's progress depended on the fullest use of the natural resources of his environment was a popular idea in the young republic."[56] It was indeed. By the nineteenth century, any fears of subordination to nature were being swiftly overturned by a drive to master it. Industrious Americans were busy chasing progress – diligently improving on what nature bestowed. The Boston Associates were very much a part of this culture, and their thoughts about the natural world are worth noting. Many of them undoubtedly perceived and valued nature as a central component of the process of production. The natural world existed as a reservoir of productive potential awaiting the contriving hand of humanity. And most important, by combining it with labor, its value to humankind rose significantly. Many of them were likely to agree with Nathan Appleton when he invoked David Ricardo's famous maxim: "Labor acts upon materials furnished by Nature; but Nature is gratuitous in her gifts, and it is only when acted on by man that her productions acquire value in his estimation."[57]

Appleton's fellow associate, Abbott Lawrence, expressed much the same view. Alongside his career as a founder of the textile city that bears his name, Lawrence served two terms as a United States congressman. In a letter to a colleague from Virginia in 1846, Lawrence wrote about the need to fully exploit the region's "matchless natural resources." He put his point of view thus:

> I have thought that the State of Virginia, with its temperate climate, variety and excellence of soil, exhaustless water-power, and exuberant mineral wealth, contains within herself more that is valuable for the uses of mankind, in these modern days, than any other State in our Union.
>
> I need not say to you that these gifts of Providence are of little consequence to your people, or to our common country, unless developed and improved for the purposes for which they were intended. . . .

---

[55] Here I echo Donald Worster, *Dust Bowl: The Southern Plains in the 1930s* (Oxford, 1979), 94–5.

[56] Arthur A. Ekirch, *The Idea of Progress in America, 1815–1860* (New York, 1951), 73.

[57] Nathan Appleton, *Labor, Its Relations in Europe and the United States Compared* (Boston, 1844), 3.

> The truth is that nature has been profuse in her gifts in behalf of your people, and you have done but little for yourselves.[58]

At the time he wrote this letter, Lawrence was making plans for the new textile city of Lawrence, Massachusetts. With both water and production in mind, he urged the economic development of Virginia's rivers. "The water-power on the James River at Richmond is unrivalled," he observed, "and it seems a great waste of natural wealth to permit it to run into the sea, having hardly touched a water-wheel."[59]

Nature had long been conceived of as a gift, as an endowment made out in the name of humankind. (The phrase "gyftes of nature," according to the *Oxford English Dictionary*, dates from 1504.) What is noteworthy about these early-nineteenth-century thoughts on nature is their production-oriented tone. To Appleton, Lawrence, and their associates (and to Ricardo and others as well), an economic logic informed their understanding of humanity and its relationship to the natural world. This understanding led them to perceive the environment with a great deal of calculation, in instrumental terms. In this view, the failure to improve nature, to incorporate it into the designs of humankind was both foolish and wasteful. Instead, to the Boston Associates the environment possessed the potential to yield great wealth if wisely handled, if used to serve primarily economic ends.

The behavior and attitudes of the Boston Associates reflect a commitment to controlling nature in the interests of industry, to managing water with an eye toward its productive value. But if an economic calculus informed the Boston Associates' understanding of the natural world, their thoughts, to be sure, were not all domination and conquest of nature for the sake of production. A closer look at their cultural activities – their world outside of business and industry – betrays as well a more benign and nurturing side to their dealings with the natural environment.

By the second quarter of the nineteenth century, Boston's elite showed considerable interest in the natural and earth sciences – in studying the land and water, its plants and animals. In 1830, a small group of Boston gentlemen, concerned with cultivating

---

[58] The text of Lawrence's letter is printed in the appendix of Hamilton Andrews Hill, *Memoir of Abbott Lawrence* (Boston, 1883), 138.

[59] Ibid., 140.

a better understanding of the natural world, formed the
Boston Society of Natural History (BSNH).[60] When the BSNH
began, recalled its president Amos Binney in 1845, virtually no
attempt had been made to study the natural history of Massachusetts – no reliable geological studies, no "general knowledge of the birds which fly about us, of the fishes which fill our
waters, of the shells of our beaches, or of the lower tribes of
animals that swarm both in air and in sea." What knowledge
existed of the natural world had come mainly from the work of
people outside New England. "The laborer in Natural History,"
Binney lamented, "worked solitary and alone . . . without the
approbation of the public mind, which, unenlightened as it was,
yielded no honor to persons occupied with such studies, but on
the contrary, regarded them as busy triflers."[61] There was, in
short, a vacuum of knowledge, a deficit the BSNH worked
diligently to redress.

In its early years, the BSNH tried to cultivate a scientific
sensibility. Naturalists could now neatly classify the creatures
inhabiting New England in the BSNH's new journal. The BSNH
welcomed interested, respectable persons to join them in their
endeavors, and many of Boston's elite responded. Among the
BSNH's patrons could be found the name of Nathan Appleton.
Not long after the founding of Lowell, Appleton began to reflect
on the distinct look of the earth's surface. He wondered particularly why the rock he found buried in the earth appeared rather
smooth, and published his thoughts on the matter in the *American Journal of Science and Arts*.[62] It is difficult at first to imagine
Appleton pondering how the earth's surface was formed. Too
arcane, it might be argued, for such an enterprising tempera-
ment. Yet Appleton's world was one struggling to come to terms
with the environment around it, and such concerns probably
attracted him to the BSNH. Moreover, Appleton was not alone
in his musings. In all likelihood, many of his fellow Boston
Associates looked to science to unpack the wonders of the natu-

---

[60] A brief history of the BSNH can be found in *The Boston Society of Natural History,
1830–1930* (Boston, 1930).

[61] Amos Binney, *Remarks Made at the Annual Meeting of the Boston Society of Natural
History, June 2, 1845* (Boston, 1845), 10.

[62] [Nathan Appleton], "Proofs that General and Powerful Currents Have Swept and
Worn the Surface of the Earth," *American Journal of Science and Arts* 11 (1826): 100–
104.

ral world. Indeed, of the seventy-seven men who Robert Dalzell considers Associates, thirty appear as patrons of the BSNH.[63]

The Boston Associates viewed nature as more than just an intellectual pursuit. Many of them also viewed the natural world as a source of leisure activity. Early in the nineteenth century, horticultural societies formed in New York and Pennsylvania, and in 1829 the Massachusetts Horticultural Society began. The Society organized yearly festivals and exhibitions while promoting tree planting, flower cultivation, and landscape gardening.[64] If the thoughts of one member are any indication, it was rather successful in encouraging these pursuits. "Never before," proclaimed Malthus A. Ward in 1831, "did gardening and rural affairs engross so large a share of common conversation, – often entirely excluding those unprofitable and acrimonious discussions on politics, and those religious controversies." But it was not, he continued, "the simple, the rude and uneducated, who derive the most exquisite gratification from a contemplation of the works of Nature."[65] If they did, they were not members of the Massachusetts Horticultural Society. Thirty-seven Boston Associates held membership in the Society during its first forty years of existence, and some, including Theodore Lyman, Jr., and George W. Pratt, had executive positions. Others, including John A. Lowell – one-time agent of the BMC in Waltham – and Samuel Appleton, were active financially, donating large sums of money.[66]

"There are few things more refreshing to the man of busi-

63 See *By-Laws of the Boston Society of Natural History With A List of Officers and Members* (Boston, 1891), 29. A patron was someone who contributed fifty dollars or more to the BSNH. In addition to the thirty, the BSNH lists as members for 1840 the following Associates: Edward Brooks, W. H. Gardiner, Samuel Lawrence, Thomas Motley, and Charles S. Storrow. See *Boston Journal of Natural History* 3 (1840–1): 505–9.

64 For the institutional history of the Society, see *History of the Massachusetts Horticultural Society, 1829–1878* (Boston, 1880); Albert E. Benson, *History of the Massachusetts Horticultural Society* (Norwood, Mass., 1929).

65 Malthus A. Ward, *An Address Pronounced Before the Massachusetts Horticultural Society, in Commemoration of Its Third Annual Festival, September 21, 1831* (Boston, 1831), 31–2. In America, flower gardening was not generally a lower-class pursuit as was true in England. See Keith Thomas, *Man and the Natural World: A History of the Modern Sensibility* (New York, 1983), 239–40. The appeal of flowers for nineteenth-century elites may have been due in part to a change in olfactory sensibilities. For a brilliant discussion of this, see Alain Corbin, *The Foul and the Fragrant: Odor and the French Social Imagination* (Cambridge, Mass., 1986), 188–95.

66 See *History of the Massachusetts Horticultural Society* (Boston, 1880), 123–4, 498–525.

ness," declared Ezra Weston at the Society's eighth anniversary in 1836, "that will so recruit the senses and charm the spirit as to step aside a moment from the confusion and anxiety of the street, and look upon the beauty and bounty of nature."[67] Those Associates who were dedicated gardeners, especially people such as Lyman and Peter C. Brooks who owned estates outside the city, may well have felt this way. But much more was happening in the minds of Boston's elite as they planted tulips and dahlias or stooped down to admire a pot of anemones. In the rhetoric of those who addressed the Society at its meetings, one finds the justification for such leisure pursuits. Horticulture, explained one member, was "conducive to moral and intellectual refinement."[68] If pursued consistently, gardening offered help in building "character," the mark of proper social development during the nineteenth century.

The diligent study of nature, another writer remarked, served as an antidote to what was "sordid and selfish" in the world.[69] In this sense, horticulture may have helped some people to counteract the materialistic aspects of their everyday lives. To those troubled by the world of business and commerce, horticulture held out the promise of redemption.[70] A variety of social forces – economic, religious, psychological –

---

[67] Ezra Weston, Jr., *An Address Delivered Before the Massachusetts Horticultural Society, at Their Eighth Anniversary, September 17, 1836* (Boston, 1836), 5–6. Although the evidence is fragmentary, workers may have at times perceived nature in a similar way. Lowell's women workers, for example, dreamt often of nature. Faced with the crowds and noise of factory and urban life, women workers yearned for the quiet solitude of the natural world. See Benita Eisler, ed., *The Lowell Offering: Writings by New England Mill Women (1840–1845)* (New York, 1977), 137–8.

Similarly, when the AMC in Manchester needed to cut down an elm tree in 1853 to clear space for an additional mill, five hundred men and women workers petitioned to stop the company. It was, in their opinion, "a beautiful and godly tree" that offered them "a connecting link between the present and the past." The petition acknowledged the company's well-manicured premises and the considerable effort made to create a pleasing industrial environment. But this particular tree may well have represented an element of nature as yet untouched by industrial design. The AMC apparently spared the tree. The petition is quoted in "New York Industrial Exposition," *British Sessional Papers* (Commons), vol. 36 (1854), 10. Also see Gutman, *Work, Culture & Society*, 29.

[68] John Lewis Russell, *A Discourse Delivered Before the Massachusetts Horticultural Society, on the Celebration of its Seventh Anniversary, September 17, 1835* (Boston, 1835), 7.

[69] George Lunt, *An Address Delivered Before the Massachusetts Horticultural Society, on the Dedication of Horticultural Hall, May 15, 1845* (Boston, 1845), 8. Also see John C. Gray, *An Address Delivered Before the Massachusetts Horticultural Society, at Their Sixth Anniversary, September 17, 1834* (Boston, 1834).

[70] See Tamara Plakins Thornton, *Cultivating Gentlemen: The Meaning of Country Life among the Boston Elite, 1785–1860* (New Haven, 1989), 161–5.

compelled a special interest in nature. But above all, the quest for moral rectitude, for honest, reputable, and upright character inspired much of this concern for cultivating flowers, plants, and trees.

Yet having noted this, a contradiction emerges: How can such benign and nurturing thoughts about New England's flora be reconciled with the far more aggressive and calculating attitude shown by the Associates toward water? The distinction should not be overstated; no neat line divided the Associates' perceptions of land and water. However, a benevolent attitude did seem to exist toward those aspects of the natural world removed from the process of production. Rivers and streams seem not to have experienced quite the same indulgence granted trees and plants. In particular, as rivers moved into the matrix of production, as they clearly did throughout much of nineteenth-century New England, a far more exacting and manipulative attitude held sway.[71] Their central place in production tended to exclude rivers for most of the century from the same type of regard expressed toward other areas of nature.

Still, much united the different attitudes shown those elements of nature involved with production and those that were not. When one speaker before the Massachusetts Horticultural Society pronounced the garden a place where "we should learn those principles of neatness and order," he articulated one such connection.[72] If anything, the gardens created were well-ordered ones, as well designed and administered perhaps as the emerging mill towns. The theme of order and control held together the Associates' thoughts about nature, their alternating aesthetic regard and more aggressive, utilitarian stance toward the environment. Indeed, the idea of control was a central, dominating cultural current during the nineteenth century. It was present as a guiding philosophy in the Associates' determination to shape the physical setting of their mill towns, in their neatly planned factories and boardinghouses. It surfaced in the tightly regimented schedule and routine they created for their workers.[73] And it emerged in the carefully

[71] Keith Thomas makes a similar argument in regard to the humanity shown animals. See Thomas, *Man and the Natural World*, 181–2.

[72] Gray, *Address*, 10; also see *History of the Massachusetts Horticultural Society* (Boston, 1880), 95.

[73] For a discussion of the Boston Associates and their commitment to organizing "total institutions," see John F. Kasson, *Civilizing the Machine: Technology and Republican Values in America, 1776–1900* (New York, 1976), 64–9.

controlled path they prescribed for water. The Associates set out to engineer a specific environment in the broadest sense – a physical setting inside and outside the factory consistent with their goals for production. It is hardly surprising to find this same theme present in their pursuits outside the factory towns.

The attempt to impose order on the land and waterscape is among the most striking and powerful features of industrial transformation. It is also a vital component of the quest to maximize productive potential, to efficiently capitalize on the earth's natural resources. The control of nature, whether for economic interest or leisure pursuit, underlay the advance of industrial capitalism. This idea united the Associates in their seemingly contrary attitudes toward land – its plants, trees, and flowers – and water.

3

COMPANY WATERS

If ever there was a period of sudden change in the Merrimack valley, it began in the 1820s and lasted to midcentury. These were the years when textile cities pushed their way into the watershed, casting their long shadows all across the region's landscape. After the founding of Lowell, the Boston Associates reached out over the waterscape to change rural into urban, to turn water into power and production. Large-scale waterpowered factories based on the Waltham–Lowell model multiplied throughout the region, at Nashua and Manchester, New Hampshire, in the 1830s, and the following decade in Lawrence, Massachusetts. By midcentury, the Boston Associates and their companies had completely reshaped the valley's waters.

The new waterscape was nicely tailored to the needs of large-scale textile production. Thirty river miles upstream from Lowell, the water at Amoskeag Falls tumbles fifty-four feet in the space of a mile. The spot is ideal for generating waterpower. Between 1837 and 1845, the Boston Associates financed a system for delivering water to the mills at Manchester, New Hampshire. During that time, a five-foot-high masonry wing dam, guard locks, and two power canals, each over five thousand feet in length, created a system of waterpower generation that operated on two levels. From a basin used for storage, water entered the upper canal, fell twenty feet as it passed to a lower canal and then emptied another thirty-four feet into the Merrimack River.[1] Although the two-level system of canals resembled the Lowell design, the Manchester system was far more streamlined. The topography at Lowell prevented the construction of a power canal that ran on a parallel course with

---

[1] C. E. Potter, *The History of Manchester, Formerly Derryfield, in New-Hampshire* (Manchester, N.H., 1856), 553–4.

the river, the ideal way for distributing the water. At Manchester, however, it was possible to build the canals reasonably straight and parallel to the river, a design that maximized both the fall and the waterpower.

From Manchester, the water flowed south another sixteen river miles before reaching the mouth of the Nashua River. The Nashua River is shorter than the Merrimack, its main branch totaling roughly thirty-one miles. Along the last part of its course, as it flows northeasterly to meet the Merrimack, it falls eighty-eight feet in its last twelve miles.[2] Although the Boston Associates were not the first to develop the waterpower, by the 1830s they gained control over the mills here. A canal three miles in length and six feet deep brought water to the Nashua Manufacturing Company's factories. Thirty-three feet of water was available for power.[3] One mile later, a canal one thousand feet long guided water to the mills of the Jackson Company.[4] Half a dozen factories took the water, combined it with cotton and labor, and produced textiles. After being used for production, the water ultimately poured out into the Merrimack.

From this point, the Merrimack's water rushed on toward Lowell, where by midcentury new additions had been made to the already intricate system of water control. Built during the 1840s to meet the expanded need for waterpower, the Northern Canal – 4,400 feet long and 100 feet wide – ran from Pawtucket Falls to the Western Canal. The canal was better designed and could carry more water than the older Pawtucket Canal. Two new underground waterways neatly tied the entire system together, completing the Merrimack River's most complex and elaborate hydraulic system.[5]

Twelve miles downstream from the Pawtucket Dam at

---

[2] Dwight Porter, "Water Power of the Merrimac River and of Rivers of Eastern Massachusetts between the Merrimac and the Blackstone" (ms. at MATH, 1899), 152.

[3] Charles J. Fox, *History of the Old Township of Dunstable* (Nashua, N.H., 1846), 200.

[4] George F. Swain, "Water-Power of Eastern New England," in U.S. Department of Interior, Census Office, *Reports on the Water-Power of the United States*, vols. 16 and 17 (Washington, D.C., 1885–7), 16:43 (hereafter cited as Swain, "Water-Power of Eastern New England"). A slightly different description of the water delivery system at Nashua can be found in John Hayward, *The New England Gazetteer Containing Descriptions of All the States, Counties and Towns in New England*, 2d ed. (Concord, N.H., 1839), unpaginated, see under Nashua, N.H.

[5] Patrick M. Malone, *Canals and Industry: Engineering in Lowell, 1821–1880* (Lowell, 1983), 9, 15.

Lowell, the Merrimack's water slammed into an imposing obstacle, a solid block of granite and cement averaging 32 feet in height, 35 feet thick at its base, and spanning a distance of 1,600 feet. Built in 1848 at an angle across the river, the massive Lawrence Dam – at the time unrivaled in size in either the United States or Europe – forced the water into a canal, roughly one mile in length and varying between 60 and 100 feet wide. The canal sheered off an oblong piece of land on the north side of the river, creating an island on which the textile mills of Lawrence would soon be built.[6] Lawrence was the final link in the Boston Associates' plans for redesigning the Merrimack River, the last major technological complex before the river's water went drifting out to sea.

By the middle of the nineteenth century, large-scale textile factories had gained extensive control over the waters of the Merrimack valley. In all, the cities of Lowell, Lawrence, and Manchester controlled the flow of the Merrimack River for roughly forty-five river miles, or close to 40 percent of its course from its beginning in Franklin, New Hampshire.[7] This allowed production to proceed on a massive scale. There were over forty Waltham–Lowell-style mills operating in the valley in 1850. The mills produced over $14 million worth of cotton textiles or 47 percent of the combined output of Massachusetts and New Hampshire. With few exceptions (the Massachusetts Manufacturing Company at Lowell ran partly on steam power), water met the energy needs of the mills.[8]

The companies in charge of this vast amount of water wealth had their own distinct relationship with nature. They handled water in a way different from much of what had gone on

---

6 This description is based on Peter M. Molloy, "Nineteenth-Century Hydropower: Design and Construction of Lawrence Dam, 1845–48," *Winterthur Portfolio* 15 (1980): 320; Swain, "Water-Power of Eastern New England," 25; Mass. Sanitary Commission, *Sanitary Survey of the Town of Lawrence* (Boston, 1850), 6–7.

7 The AMC alone had control of the water for twenty river miles north and south of Amoskeag Falls. The Pawtucket Dam at Lowell controlled the water for eighteen miles upstream. The Lawrence Dam could be used to raise the water legally in the Merrimack to the foot of Hunt's Falls, about seven river miles upstream.

8 The figures for the four valley towns were tabulated from the Manuscript Schedules, U.S. Census of Industry, 1850. The figure for combined output in Massachusetts and New Hampshire is based on the computed value of product reported in 1850 equal to roughly $30.3 million. This figure is from U.S. Congress, *Manufactures in the Several States and Territories For the Year Ending June 1, 1850: Abstract of the Statistics of Manufacturing According to the Returns of the Seventh Census*, 35th Cong., 2d sess., Senate Exec. Doc. 39, 43.

before, developing ingenious and novel methods for selling water, for conceiving of it as a form of property. Managing water in the interests of production presented, as well, a number of technical difficulties. For solutions, the corporations turned to men of science, to some of the most talented civil engineers in America. Invoking science to help them master the natural world, the companies employed such men to further regulate and rationalize the use of water.

Before the Boston Associates and their corporations, disparate groups of people, often local lawyers and merchants, developed the substantial water resources of the Merrimack valley. Daniel Abbot, a Harvard-educated lawyer born in the valley town of Andover, Massachusetts, and his law partner Benjamin French were the original promoters of Dunstable (later the city of Nashua) as a site for textile manufacturing. In June 1823, the state of New Hampshire granted a charter for the Nashua Manufacturing Company to Abbot, a tavern owner named Moses Tyler, and Joseph Greeley, the proprietor of a general store in Dunstable.[9] Likewise, the idea for first developing the waterpower near where the city of Lawrence would eventually emerge was that of J. Gardiner Abbott, a lawyer living in Lowell, and his uncle Daniel Saunders, who practiced law in nearby Methuen.[10] Samuel Slater of Rhode Island, his son-in-law Learned Pitcher, the Boston Associate Willard Sayles, his business partner Lyman Tiffany, and Oliver Dean, a Massachusetts cotton manufacturer, were behind the first serious attempt to make use of the water at Amoskeag Falls (Manchester).[11]

None of these ventures was terribly successful. Neither the Nashua Manufacturing Company nor its companion, the Indian Head Company, located at the lower mill site on the Nashua River, saw a great deal of financial success. Undercapitalized from the start, the Nashua Company did not return a dividend until 1828, and the Indian Head Company was continually

[9] Elizabeth MacGill, "One Hundred Years of Cotton Textile Manufacture Rounded Out by Nashua's Major Industrial Concern," *Nashua Telegraph*, 24 June 1924.

[10] Duncan Erroll Hay, "Building 'The New City on the Merrimack': The Essex Company and its Role in the Creation of Lawrence, Massachusetts" (Ph.D. diss., University of Delaware, 1986), 37.

[11] Robert V. Spalding, "The Boston Mercantile Community and the Promotion of the Textile Industry in New England, 1813–1860" (Ph.D. diss., Yale University, 1963), 133.

short of funds.[12] Daniel Abbot at first envisioned eight factories sprawled across the Nashua site. But in 1828, the Nashua Manufacturing Company had just two factories in operation with close to eleven thousand spindles combined. Still, the company had made significant progress in developing the available waterpower. Abbot hired James F. Baldwin and a millwright named William Boardman to oversee construction of the dams and canals.[13] The youngest son of Loammi Baldwin, Sr. – the chief engineer of the Middlesex Canal – James Baldwin guided the construction of a three-mile-long canal that brought water from Mine Falls to the Nashua Manufacturing Company's mills. In 1825, the company's first factory began partial operation, and work began on another dam and canal. When completed, this canal created a second, artificial passage for water between the Nashua and Merrimack rivers (a natural passage of course already existed between the two rivers). Earlier, the Nashua Company had decided to develop and market the waterpower of this lower privilege, the site eventually occupied by the Indian Head Company in 1826.[14]

The disappointing financial fortunes of the mills at Dunstable were not all due to organizational problems and financial mismanagement. They also stemmed from technical and environmental difficulties. In particular, attempts to control the flow of water proved problematic. As early as 1825, the wasteway of the upper canal failed to work properly, and fears surfaced that the canal itself would at some point give way. The sandy soil found in the area made it difficult to build reliable and sturdy canal walls, a point underscored by a breach in the lower canal that opened one year later. In 1827, torrential spring rains damaged the lower dam causing its apron (that part of the dam built on its downstream side to keep flowing water from undermining the riverbed) to break away and the center ridge to settle. Then the Nashua and Indian Head companies clashed over who would assume the costs for repairing the destruction at the lower mill site – a dispute finally settled in 1828.[15] By the following year, the stock of the Indian Head

[12] Ibid., 120, 126.

[13] D. Abbot, Statement of the Present Condition & Prospect of the Operation, 2 June 1824, box H-1, NMC Papers, BL; Annual Report, 6 June 1827, and Treasurer's Report, 28 May 1828, vol. AB-1, ibid.

[14] MacGill, "Cotton Textile Manufacture"; Directors' Reports, 16 Mar. and 2 Sept. 1825, vol. H-1, NMC Papers.

[15] Directors' Reports, 4 Jan. 1825 and 16 Mar. 1825, vol. H-1, NMC Papers; Report of

Company had declined sharply. The Boston Associates, led by William and Nathan Appleton, Eben Francis, and David Sears, began buying up devalued shares of the Indian Head stock. They took control of the mills in 1830, reincorporating under the name of the Jackson Manufacturing Company.[16]

Several years later, the Boston Associates again set out to control more land and waterpower, this time at Amoskeag Falls in Manchester. In April 1836, Oliver Dean and his colleagues stepped aside as directors of the Amoskeag Manufacturing Company, although they continued to hold stock. Six new directors replaced them, four of whom were Boston Associates – Patrick Tracy Jackson, George Bond, Samuel Frothingham, and William Appleton, Nathan Appleton's cousin. Francis Cabot Lowell, Jr., the son of the founder of the Boston Manufacturing Company, was appointed treasurer. And as the capital of the company expanded between 1835 and 1838, from $200,000 to $1 million, the names of prominent Associates, including Nathan Appleton and Abbott Lawrence, appeared as stockholders.[17]

Meanwhile, in the summer of 1835, the Amoskeag Company began an aggressive attempt to control all the waterpower from Sewall's Falls in Concord (twenty-six miles north of Amoskeag Falls) to several miles past the southern limit of Manchester. The company first acquired the stock of the Hooksett Manufacturing Company – a textile mill of roughly seven thousand spindles – giving it possession of the waterpower at Hooksett Falls, about seven miles above Manchester. During the following fall and early winter, the Amoskeag Company purchased three canal companies – the Union, Isle of Hooksett, and Bow Canal – a move that gave it effective control over the streamflow for approximately twenty miles north and south of Amoskeag Falls. In the spring of 1836, the company acquired additional land and water rights at Garvin's Falls when it absorbed the stock of the Concord Manufacturing Company. Finally, the company already owned a significant chunk of property in Goffstown, on the western side of Amoskeag Falls, which it maintained to avoid any interference with its management of

Company Agent, 7 June 1826, vol. K-1, Papers of the Indian Head Factories, BL; Annual Report, 6 June 1827, vol. AB-1, NMC Papers; Agreement Signed Between NMC and Indian Head Factories, 1 Nov. 1828, vol. K-1, Papers of the Indian Head Factories.
16 Spalding, "Boston Mercantile Community," 126–7.
17 Ibid., 139.

the waterpower.[18] In but several months' time, the Amoskeag Company – now controlled by the Boston Associates – had grown from a small-scale producer of textiles into a veritable hydraulic empire encompassing a vast expanse of land and water in the central Merrimack valley.

In the following decade, the Boston Associates consolidated their control over the lower Merrimack River. The move to assemble land and water rights at Bodwell's Falls downstream from Lowell was largely the work of a single person, Daniel Saunders, who was not an Associate. Together with his son, Daniel Saunders, Jr., J. G. Abbott, Samuel Lawrence, a Boston Associate, John Nesmith, an inventor and manufacturer living in Lowell, and others, Saunders formed the Merrimack Water Power Association (MWPA) to purchase land along the river.[19] Saunders had title (or conditional title) by 1845 to several square miles of land along the Merrimack, from Hunt's Falls (below Lowell) downstream to the mouth of the Shawsheen River.[20]

Prospects for another major textile venture looked dim in the early 1840s. A serious downturn in the economy occurred during this time, and profits in New England's textile industry suffered severely. But profits, which bottomed out in 1843 at 2.3 percent of net worth, recovered and shot up during 1844, reaching over 19 percent according to a study done of eleven New England textile firms (five Lowell companies were included).[21] The following year, the Boston Associates founded the Essex Company. Spearheaded by Abbott Lawrence, this waterpower company, which had the authority to erect a dam across the Merrimack, purchased the holdings of the MWPA for $30,000.[22] The principal investors in Lawrence, Massachusetts, were Abbott and Samuel Lawrence, Nathan Appleton, John Amory Lowell, Thomas H. Perkins, George W. Lyman, and J. Wiley Edmunds.[23] Having secured control of the

---

18 George Waldo Browne, *The Amoskeag Manufacturing Co. of Manchester, New Hampshire: A History* (Manchester, 1915), 55–6; Spalding, "Boston Mercantile Community," 138–9; E. A. Straw Engineering Notebook, 1875–8, location D-2, AMC Papers, MHA.

19 H. A. Wadsworth, *History of Lawrence, Massachusetts* (Lawrence, 1880), 44.

20 Hay, "Building 'The New City on the Merrimack,'" 42.

21 Robert F. Dalzell, Jr., *Enterprising Elite: The Boston Associates and the World They Made* (Cambridge, Mass., 1987), 52.

22 Hay, "Building 'The New City on the Merrimack,'" 44–7.

23 Spalding, "Boston Mercantile Community," 180.

water at Bodwell's Falls, the Boston Associates now claimed a total of four waterpower sites in the Merrimack valley.

The Boston Associates organized corporations with large amounts of capital to carry out their plans to produce textiles. By the mid-1840s, the ten existing Lowell companies had an average capitalization of slightly over $1 million; all the textile corporations at Lawrence were capitalized at more than this figure. (By contrast, the original capitalization of the BMC in Waltham was $400,000.)[24] Moreover, the stock of the textile companies remained in the hands of the Boston Associates and their families. These were closely held ventures, and nothing akin to a modern-day public offering of stock ever took place. The mills and their workers were typically managed by an agent who supervised the daily operation of the factories. A treasurer directed the company's financial planning, bought the necessary raw materials, and sold the finished goods.[25]

More specialized corporations, distinct from the ones that produced textiles, took care of the waterpower needs of the textile mills. Such corporations, formed to sell and manage water, had not existed before the nineteenth century. The Merrimack valley waterpower companies were new and innovative enterprises with ambitious plans for production, and with corresponding strategies for making the environment cooperate. The PLC, the Amoskeag Company in Manchester, and the Essex Company in Lawrence (no separate waterpower company was formed in Nashua) were all established to create waterpower and oversee its distribution and regulation.

The PLC in Lowell was the prototype for nineteenth-century waterpower companies. Founded in 1792 to build a canal around Pawtucket Falls in Chelmsford, the PLC was acquired by the Boston Associates in the early 1820s. The Pawtucket Canal and the nearby land and water rights were later transferred to the Merrimack Manufacturing Company. By 1826, however, the Merrimack Company shifted ownership of the canal, the land, and water privileges back to the PLC and focused its energies exclusively on the production of textiles. The PLC operated a machine shop and presided over the creation, distribution, and management of waterpower.[26]

24 Henry A. Miles, *Lowell, As It Was, and As It Is* (Lowell, 1845), 48–57; Dalzell, *Enterprising Elite*, 26, 71.
25 Dalzell, *Enterprising Elite*, 49–50, 56–8.
26 Louis C. Hunter, *Waterpower in the Century of the Steam Engine*, A History of Industrial Power in the United States, 1780–1930, vol. 1 (Charlottesville, Va., 1979), 212–13.

The PLC and the other waterpower corporations that followed forged their own special relationship with nature. In particular, the Waltham–Lowell waterpower companies diverged from the established customs for the sale of land and water rights. In eighteenth-century New England, riparian landowners connected ownership of the streambank with rights (not ownership) to the water at that point. Riparian land and water privileges were inseparable, and the two generally were sold together. In the technical legal sense, water was *appurtenant* to land. This meant that water was deemed incidental to land, giving the property owner the right to use it in conjunction with the land itself. When mills changed hands during the eighteenth century, deeds might state that the owner was conveying with the mill buildings and land that "part of the Stream as may be necessary for the Improvement of said Works."[27] It was assumed that landowners had the right to make use of the water flowing through their land. No precise figures were generally given for the quantity of water conferred along with the land.

When the PLC sold the water at Lowell it did so in a completely different manner. The Merrimack Manufacturing Company was the first textile firm to purchase land and water from the newly organized waterpower company. For a sum of $40,264, the PLC conveyed thirty acres of land and seven "mill privileges" according to a deed signed by the two companies in 1826.[28] Unlike earlier deeds regarding water rights, this one expressed the PLC's intention to sell discrete quantities of water. The form of this deed, in use through the 1830s, explicitly stipulated the PLC's terms for the sale of waterpower.[29] Most important, the company gave the term mill privilege, which had long been used in deeds, a completely different meaning. In particular, the company introduced the *mill-power* concept – a term the PLC used interchangeably with mill privilege. The term was coined, appropriately enough, in Waltham. It originally equalled the power required to run 3,584 spindles

---

27 The quote is from J. Hubbard to N. Hubbard, 12 Jan. 1705, bk. 15:143, Middlesex County Registry of Deeds. The property involved was located at Newton Lower Falls on the Charles River.

28 PLC to MMC, 8 June 1826, bk. 6:230–5, Middlesex County Registry of Deeds, Northern District, Lowell, Mass.

29 See Proposals & c. Annexed to and Made Part of the Indenture Hereafter Recorded at Page 230 & c., 8 June 1826, bk. 6:220–30, Middlesex County Registry of Deeds, Northern District. This proposal was generally annexed to all deeds for the sale of waterpower and land at Lowell through the 1830s.

in addition to all the preparatory equipment and looms necessary to weave the yarn into cloth – the capacity of the Boston
Manufacturing Company's second integrated cotton mill. Since
different kinds of machinery used various amounts of power, a
mill power was further standardized in the PLC's deed to equal
a streamflow of 25 cfs at a fall of thirty feet. At Lowell, where a
two-level canal system existed, mills with thirteen and seventeen feet of fall were entitled, respectively, to 60.5 and 45.5 cfs
of streamflow.[30]

The term mill power is, in a sense, a misnomer. It is not a unit
of power in the conventional sense that power measures the
timed rate at which work is performed.[31] Instead it signified a
discrete quantity of water available for purchase from the PLC.
The other Merrimack valley waterpower companies eventually
adopted the mill-power concept, as did companies in other valleys in New England where mills on the Waltham–Lowell
model emerged. The development of the mill power represented a convenient way of bundling water and putting it up
for sale. Its significance as a measurement cannot be divorced
from the intention to market the water. The term connoted, as
well, water's productive potential – it being first conceived as
enough streamflow to operate one large, integrated textile factory. As a practical matter, the PLC sold water at a rate that
varied from three to four dollars per spindle, the cost of a
single mill power ranging from $10,752 to $14,336.[32] By 1846,
the PLC had sold slightly over 91 mill powers; by the following
decade the number had increased to 139.[33]

Technically, the PLC never outright *sold* mill powers to the
textile companies at Lowell. Instead, the deeds arranged for
the mills to pay an annual rent of three hundred dollars to the
waterpower company (the mills typically withheld five thousand dollars of the sum they agreed to pay for the mill powers
and used the interest on this amount to cover the annual rental
cost). The arrangement required the PLC to maintain the
Lowell dams and canals in good condition. If the company

---

30 J. Francis to B. Saunders, 21 Oct. 1864, vol. DB-3:565–71, PLC Papers, BL. Also see
Proposals, 8 June 1826, bk. 6:220.
31 Hay, "Building 'The New City on the Merrimack,'" 468.
32 J. Francis to A. Lawrence, 2 July 1859, vol. DB-2, PLC Papers, and J. Francis to B.
Saunders, 21 Oct. 1864.
33 The mill-power figures were calculated to represent the minimum available year-
round flow. Directors' Records, 24 Apr. 1849, vol. 3, PLC Papers; J. Francis to A.
Lawrence, 2 July 1859.

failed to do so, the affected textile mills could make the repairs themselves, withholding rent during this period.[34]

In a stunning break with the past, the PLC severed water from land, making possible the eventual separate sale of both elements of the natural world. During the 1820s, small pieces of land were conveyed to the Lowell mills along with the right to use the water. But the language of the deeds signed by the PLC and the Lowell mills suggests a radically new understanding of the relationship between land and water. A remarkable statement found in the PLC's deeds notes that "where several mill powers are granted in the same deed . . . the land conveyed with them is not therein specially divided appointed and attached to each."[35] Such language implied that rights to land and water were considered apart from one another, opening the way for the separate sale of water. By the 1830s, the PLC sold water without including any land at all – a pivotal development in the commodification of this resource.[36]

Although land sometimes accompanied the sale of waterpower at Lowell and other Merrimack valley cities, the land itself did not give a textile company the right to use all the water at that point. Rather, the permissible amount of water was stipulated by the arbitrary definition of a mill power. Individual corporations purchased the right to draw a specific amount of water to drive their mills and were legally entitled to no more.

At small, privately owned mills, most deeds written in colonial America referred only to the sale of mill privileges or shares in them. In contrast to the waterpower companies formed by the Boston Associates, no reference was made to the quantity of water to which the new owner was entitled. Of course, the sale of mill privileges in early America entailed, by definition, exchange. But by the nineteenth century, the parameters of exchange were changing to reduce water ever more thoroughly to an abstraction. The Merrimack valley waterpower companies sold waterpower packages that promoted industry in these planned industrial cities. Once the power was purchased, the

---

[34] J. Francis to B. Saunders, 21 Oct. 1864. Also see Proposals, 8 June 1826, bk. 6:220. To guard against currency depreciation, the PLC stipulated that rent be paid in so many ounces of gold or silver.

[35] Proposals, 8 June 1826, bk. 6:228.

[36] See, e.g., PLC to MMC, 28 Sept. 1832, bk. 13:339, and 1 Sept. 1832, bk. 18:373; PLC to Hamilton Manufacturing Company, 10 Aug. 1833, bk. 17:303; PLC to Appleton Manufacturing Company, 10 Aug. 1833, bk. 17:319, Middlesex County Registry of Deeds, Northern District.

waterpower companies were generally hired to build the mills
and the necessary machinery for textile production. At Lowell,
the PLC earned large profits from the sale of textile machinery,
and probably sold waterpower there with this fact in mind.
According to one source, little if any profit was earned on the
actual sale of waterpower.[37] However little was earned, the
marketing of waterpower at Lowell and the other Merrimack
valley factory towns embodied a relationship with nature that
defined water principally in terms of its value for exchange and
production.

The commodification of water coincided with the increasing
rationalization of this resource. To make the most productive
use of nature, to obtain whatever last drop of waterpower potential the Merrimack had to offer, the valley waterpower companies relied on the scientific expertise of a number of civil
engineers.

Loammi Baldwin, Jr., was instrumental in the early development of the Waltham–Lowell-style mills. A graduate of Harvard
at the age of twenty, Baldwin, the older brother of James and the
most distinguished member of this family of civil engineers, had
turned to engineering after a brief career as a lawyer. In his
thirty years as an engineer, Baldwin undertook a range of projects, including the design of the Bunker Hill Monument and the
dry docks at the Charlestown, Massachusetts, and Norfolk, Virginia, navy yards. Yet throughout his career, he remained mostly
concerned with water. He redesigned a long list of rivers, surveying, building canals and waterpower systems all across the eastern half of America – on the Schuylkill, Altamaha, Androscoggin, Kennebec, Concord, Charles, and Merrimack rivers. He
had, according to one biographer, "the power of enlisting
Nature as an aid and of turning her forces into profitable
channels."[38]

Baldwin was a temperamental sort. In 1821, he had been
hired to work on the Union Canal in Pennsylvania, but in a
space of two years had twice resigned from the project, before
finally being fired by the management. Baldwin's contempt ran
high after the company repeatedly challenged his control over

[37] J. Francis to B. Saunders, 21 Oct. 1864. On the affairs of the PLC, see George S.
Gibb, *The Saco-Lowell Shops: Textile Machinery Building in New England, 1813–1949*
(Cambridge, Mass., 1950), ch. 3.
[38] George L. Vose, *A Sketch of the Life and Works of Loammi Baldwin* (Boston, 1885), 6, 27.

the work. When the company eventually hired two outside engineers to give additional opinions on the route the canal should take, Baldwin complained incessantly and insisted that their appointments should not have been made without his approval. To the end, Baldwin defended his authority as a self-reliant professional. He refused to bow to any interests outside the sphere of science — financial or otherwise. Vehemently independent, he wrote to his brother James at one point that he could not "bear to be *managed.*"[39] The unmanageable manager, Baldwin considered himself a rational man dedicated to the pursuit of the higher ideals of science. Unlike the millers and mechanics of his day, who he charged knew nothing of the "natural & subtle principles of hydraulics," he believed in the value of reliable, scientific knowledge.[40]

In the summer of 1836, Kirk Boott, Patrick Tracy Jackson, and Oliver Dean conferred with Francis Cabot Lowell, Jr., on how best to take advantage of the waterpower at Amoskeag Falls.[41] Despite such experienced hands, Francis Lowell sought Baldwin's advice to help them place the water there under industrial control. That same summer, Baldwin visited the Amoskeag site. The place was not entirely unknown to Baldwin or his family. His father had played a part in surveying the original Blodget Canal (renamed the Amoskeag Canal), which ran along the eastern side of the falls.

Writing Francis Lowell following his visit that summer, Baldwin mentioned the lack of scientific foresight applied earlier in the century at Lowell, Massachusetts. Raising the height of the Pawtucket Dam by two feet during the early 1830s (see p. 65) had injured the Jackson Company's mills upstream in Nashua. The dam now flowed the Merrimack's water all the way to the Nashua River, impeding the company's waterwheels.[42] Refer-

[39] Quoted in Daniel Hovey Calhoun, *The American Civil Engineer: Origins and Conflict* (Cambridge, Mass., 1960), 98.

[40] L. Baldwin to W. W. Ellsworth, 4 June 1832, vol. 16, file 5, Baldwin Collection, BL; also see Calhoun, *American Civil Engineer,* 94–9.

[41] Directors' Records, 24 Aug. 1836, vol. A-2, AMC Papers, BL.

[42] Baldwin made this discovery as he surveyed the river for the Jackson Company. The company learned in 1832 that the dam at Pawtucket Falls would be raised two feet higher. William Amory, the company's treasurer, believed that the Pawtucket Dam at Lowell (made permanent in 1830) had caused the water in the Merrimack to rise, at some points over four feet higher than what it was before. The change in streamflow backed water onto the Jackson Company's waterwheels (which had recently been raised a foot higher) and impeded them from turning. See W. Amory to D. Sears, 1 Dec. 1832, vol. L-1, JMC Papers, BL.

In 1834, Baldwin reported that the mills at Nashua were injured by the Pawtucket

ring to this unfortunate development, Baldwin wrote Francis Lowell that

> I should not wish to see another great work on the same river suffer from the want of attention to scientific principles. . . . Had their [the PLC's] canal from the river been constructed on a proper form, they might have had more water about 2 feet higher, & not have been obliged to raise their dam to get what they now obtain.[43]

In Baldwin's mind, the problem at Lowell was clear: Unscientific principles were employed during the construction of the water delivery system – a fact with unfortunate consequences further up the river. Too little scientific rigor for Baldwin's taste had been applied there. But more was at stake here. An important assumption lay embedded in such thoughts, a vision that science could deliver more from the natural world, that it could help maximize the gain to industry.

Baldwin hoped that history would not repeat itself, that ill-considered plans would not spoil the remarkable waterpower at Manchester. Yet the state of hydrological science in the early nineteenth century made it almost certain that errors of this sort would be made. At the time, there were few reliable measures of streamflow available for accurately constructing waterpower systems.[44] Nonetheless, Baldwin did give the Amoskeag Company some extremely useful advice. He advised the company to build its factory town on the east bank of the river. That way the mills would be able to make use of the largest portion of the available waterpower. Baldwin believed the east side of the river had several advantages: The water could be more easily directed and drawn to that side, the land was level and well-situated for the construction of the mills and boarding-houses, and since the Amoskeag Canal already existed there, it seemed an obvious choice.[45] By choosing to accept Baldwin's counsel and locating its mills on the east bank of the Merrimack, the company cleared the way for large-scale production, for the use of water on a systematic and comprehensive scale.

Dam. The following year, the PLC and the Jackson Company agreed to a cash settlement. See L. Baldwin to W. Amory, 1 Nov. 1834, vol. 5, file 13, Baldwin Collection; W. Amory to PLC, 17 Mar. 1835, vol. L-1, and W. Amory to J. A. Lowell, 28 Apr. 1835, vol. L-2, JMC Papers.

[43] "Report on Amoskeag Manufacturing Company's Water," L. Baldwin to F. C. Lowell, 29 Aug. 1836, vol. 6, file 5, Baldwin Collection.

[44] Robert B. Gordon, "Hydrological Science and the Development of Waterpower for Manufacturing," *Technology and Culture* 26 (1985): 206.

[45] "Report on Amoskeag," 29 Aug. 1836.

With Baldwin dead by 1838, the founders of the Essex Company at Lawrence turned to Charles S. Storrow, a second important Merrimack valley engineer. A Harvard graduate, Storrow had spent his senior year foraging through the books and papers in Baldwin's library – one of the finest engineering libraries in the nation – having been introduced to the well-known engineer by his father. Off to Paris in 1829 to study civil engineering at the Ecole nationale des ponts et chaussées, Storrow returned three years later to begin work on the Boston & Lowell Railroad.[46] But water, not railroads, was to be Storrow's consuming passion.

In 1835, at just twenty-six years of age, he published *A Treatise on Water-Works*, a book that reviewed the prevailing theory for conducting water through pipes and channels. His chief concern was not with water as a source of power, but as a much needed resource for the "health, cleanliness, and comfort" of large, crowded cities. Much of the book is devoted to explaining the theory behind flowing water, to discussing the progress of the evolving science of hydraulics. To its core, the book reflects Storrow's firm confidence in the empirical value of science for advancing the cause of civilization. "At the present day," he explained, "when we see so many ways of applying the wealth and resources of civilized communities to the amelioration of their condition . . . it is of the greatest importance not to waste in rash and hasty enterprises those means, which, employed in some other manner, might have been productive of happy effects."[47] Like Baldwin, Storrow too believed in casting a rational eye to the question of how to best use the world's resources. And to this end, the measured and precise world of science seemed to him the most appropriate path to progress.

"The life of an engineer I have always supposed to be a laborious one," Storrow wrote Baldwin in 1829, "as he is liable at any time to be called from one part of the country to another." But, asked Storrow, "Are there, at the same time, no prospects of permanent settlement?"[48] Perhaps not at that time, but fifteen years later Storrow would have his wish for

[46] Allen Johnson and Dumas Malone, eds., *Dictionary of American Biography*, 22 vols. (New York, 1936–58), 9:98.
[47] Charles S. Storrow, *A Treatise on Water-Works for Conveying and Distributing Supplies of Water* (Boston, 1835), 1, 3.
[48] C. S. Storrow to L. Baldwin, 4 Mar. 1829, Baldwin Collection, Department of Special Collections, University of Chicago Library.

steady employment. For forty-four years, beginning in 1845, Charles Storrow dedicated his life to Lawrence, Massachusetts. He masterminded the city, the Essex Company vesting him with responsibility for planning the location of buildings, streets, and bridges, and for designing the industrial complex, including the dam and power canals. At Bodwell's Falls, the Merrimack dips a mere five feet, in contrast to a decline of six times that at Pawtucket Falls, and more than ten times that amount at Amoskeag Falls. A vast structure was needed to generate sufficient water to supply the mills Storrow anticipated would be built on the northern shore of the Merrimack River. A monument to Storrow's technical genius, the Lawrence Dam, which cost $250,000 to build, created roughly twenty-eight feet of water for supplying factories with energy.[49] With help from Charles Bigelow, a former captain in the United States Army Corps of Engineers, and an accomplished crew of junior engineers and stonemasons, Storrow spent three years overseeing construction of the dam, which created a pond of over nine thousand acre-feet behind it[50] (Figure 3.1).

Along with Charles Storrow, James Bicheno Francis ranked among the most talented and innovative of the valley's engineers. More than anyone, Francis was responsible for improving the efficiency of Lowell's waterpower system. Francis was born in the British town of Southleigh in 1815. His adolescent years were spent helping his father construct a canal and harbor for a railroad in southern Wales; he later worked for the Great Western Canal Company. In 1833, Francis emigrated to the United States and was hired by George W. Whistler, a prominent New England engineer known for his work with locomotives, who the following year became the chief engineer of the PLC. Three years later, Whistler resigned his post and Francis replaced him, marking the beginning of a career that spanned forty-five years as Lowell's foremost civil engineer.[51]

In his years as chief engineer, Francis served as a consultant to all the companies at Lowell on waterpower matters. By the 1840s, the mills there were expanding their manufacturing operations beyond the capacities of the Merrimack to supply waterpower. A technical difficulty – the loss of power due to the

[49] Mass. Sanitary Commission, *Town of Lawrence*, 7.
[50] Molloy, "Nineteenth-Century Hydropower," 319–22, 330.
[51] Johnson and Malone, *Dictionary of American Biography*, 3:578–9; William E. Worthen, "Life and Works of James B. Francis," *ORHA* 5 (1894): 227–42.

Figure 3.1. Workers building the Lawrence Dam in the summer of
1847. Painting by Marshall Field. (Courtesy of the Museum of American Textile History.)

friction of water in the canals – exacerbated the problem. As
the demand for water increased, the current in the canals rose
to accommodate the demand for power, increasing the friction,
and thereby reducing the available power.[52] Ultimately, it fell to
Francis to see that the mills were adequately supplied with
water. His first important task as chief engineer came in 1840
when he was asked to measure the flow of the river over time.
His calculations suggested that holding back the water for nine
hours during the night, when the mills were shut down, would
increase the number of available mill powers.[53] From this point
on, corporations leasing waterpower were entitled to the flow
for fifteen hours per day – a move that further rationalized the
use of water.

Francis also played an important role in overseeing the transition from waterwheel to turbine technology.[54] Until the latter

[52] Malone, *Canals and Industry,* 6, 9.
[53] Directors' Records, 12 Sept. 1840, vol. 2, PLC Papers.
[54] Hunter, *Waterpower,* 335–8.

part of the 1840s, most Waltham–Lowell mills relied on breast wheels for their power. A type of vertical waterwheel, the breast wheel required water to be let into its buckets on the upstream side, its name referring to the apron that held the water on the wheel for part of its rotation. The wheel relied on both weight and impulse to deliver its power and worked at roughly 60 percent efficiency. In 1844, Uriah Boyden – a self-educated engineer and inventor who helped design Manchester's hydraulic system – installed a turbine in the Appleton mills at Lowell. Unlike the breast wheel, water moved through the horizontally situated turbine, reacting against the curved shape of its buckets. Turbines possessed a number of distinct advantages over breast wheels. They were about 75 percent efficient, more compactly engineered than vertical wheels, and cast in iron because of their sophisticated design (wood lacked the necessary precision). And since they were submerged, backwater was not a problem.[55]

Neither Francis nor Boyden had any formal education as an engineer. Yet together they collaborated on what came to be known as the mixed-flow Francis turbine, a device that took this name because the water flowed both inward and down. Both men were hostile to mathematical theory, and emphasized, instead, a reliance on experimental research. They were, in short, creative, practical-minded men who employed empirical science to develop a more efficient technology for controlling water.[56] Their primary goals were not to contribute per se to scientific knowledge. Rather, they sought to make science more functional, to tailor it more effectively to the engineering questions at Lowell and help the textile companies to capitalize on their use of water. They were remarkably successful, so triumphant in their endeavors that by the early 1850s, they made Lowell a center of hydraulic innovation.

Over the course of his tenure as engineer, Francis helped develop a system for comprehensively manipulating the waters of the Merrimack – a lifetime dedicated to the control of water. He is best known for supervising the construction of Lowell's Northern Canal, a massive project that cost over a half-million

---

[55] James B. Francis, *Lowell Hydraulic Experiments* (Boston, 1855), 1–2; Hunter, *Waterpower*, 302, 321, 330.

[56] Edward T. Layton, Jr., "Scientific Technology, 1845–1900: The Hydraulic Turbine and the Origins of American Industrial Research," *Technology and Culture* 20 (1979): 70–3.

dollars and eventually supplied water to the Tremont, Lawrence, and Suffolk mills.[57] His efforts at Lowell left a lasting impression on his colleague William Worthen. As he put it after Francis's death in 1892: "He prepared the rules and regulations, designed the appliances for the measurement of water in the canals, organized an engineer force [*sic*] for the work . . . with the result that now for fifty years the vast body of water has been judiciously, economically, and satisfactorily distributed."[58]

In the early 1850s when Francis began writing what became his best-known treatise, *Lowell Hydraulic Experiments,* he noted that the amount of waterpower developed in the Merrimack valley far exceeded the total for all of France.[59] Industrialization had come rather suddenly to the Merrimack valley and with it, a vast change in the way its waters were handled. With the rise of the Waltham–Lowell system, there evolved sprawling waterpower infrastructures that mastered the water resources of the region. Here in this valley emerged a set of systems for controlling water on a scale without parallel before the nineteenth century. The valley's waters drew together a small group of entrepreneurs, agents, and engineers with enterprising plans for production. Compelled by these plans to control the natural world, they developed the water, improved it for sale, and managed it with an eye toward its economic potential. But the handling of water in this way divided as much as it united. The intrusion of industrial capital operated to the advantage of some but not to all. And as the century wore on and the flow came increasingly into the hands of the Boston Associates, there were others living here who dared to challenge, who rose up against the Associates' designs for the valley.

[57] See Malone, *Canals and Industry,* 15.
[58] Worthen, "James B. Francis," 234.
[59] Francis, *Lowell Hydraulic Experiments,* ix.

Part II

MATURATION

4

THE STRUGGLE OVER WATER

In 1859, the United States was on the verge of war. And so were the Merrimack valley waterpower companies controlled by the Boston Associates.

The roots of that conflict dated to 1845 when the Associates struck out after New Hampshire's largest lakes. As part of the plan, the waterpower companies at Lowell and Lawrence, the PLC and the Essex Company, each purchased half the shares of stock in a third company, the Winnepissiogee Lake Cotton and Woolen Manufacturing Company (the Lake Company) of New Hampshire.[1] Acting on behalf of the factories at Lowell and Lawrence, the Lake Company gained control over New Hampshire's four largest freshwater lakes, over water that ultimately made its way downstream to the mills along the lower Merrimack River.[2] The Lake Company and its agent in 1859, Josiah French, managed the lakes to help the textile factories in Massachusetts meet their needs for waterpower, especially during the dry summer months.

The trouble began in August 1859 when James Worster and George W. Young complained to French that the Lake Village Dam had overflowed their property.[3] The Lake Company's dam at Folsom's Falls in Lake Village raised the water level in both Paugus Bay and Lake Winnipesaukee – at times flooding land lying nearby – to store water for the mills downstream in Massachusetts (Map 4.1). Planning to be away from the area, French felt troubled enough by Worster and Young to write

---

[1] The spelling of the company's name varies. I have retained the exact spelling used in all documents cited in these notes.

[2] New Hampshire's largest freshwater lakes in descending order of size are Lake Winnipesaukee, Squam Lake, Lake Winnisquam, and Newfound Lake. Umbagog Lake, which has an area bigger than Squam Lake, is partly in Maine.

[3] J. French to J. Francis, 3, 5 Aug. 1859, vol. A-38, file 205, PLC Papers, BL. No discrete set of records exists for the Lake Company. The company's papers are included in the PLC Papers and in the EC Papers, MATH.

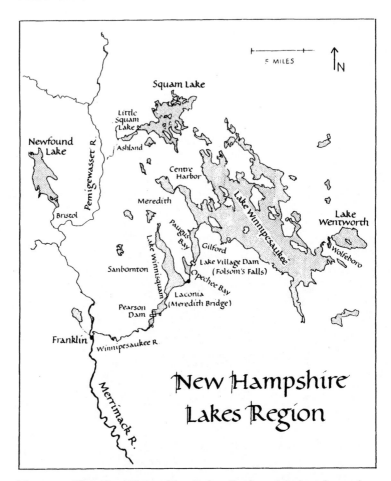

Map 4.1. The New Hampshire Lakes Region showing the major lakes feeding the Merrimack River. Except for Lake Village (now named Lakeport), all the other names are the modern spellings. The nineteenth-century names for Lake Winnipesaukee, Lake Wentworth, Paugus Bay, Opechee Bay, and Lake Winnisquam were, respectively, Winnepiseogee Lake, Smiths Pond, Long Bay, Round Bay, and Great Bay or Sanbornton Bay. (Drawing by David R. Forman.)

James Francis, chief engineer of the PLC, about finding "two or three men that would be sure and reliable in case of trouble . . . men that have the right grit."[4]

George Young, brandishing an iron bar, tried to damage the

[4] J. French to J. Francis, 11 Aug. 1859, vol. A-38, file 205, PLC Papers.

company's dam at Lake Village the following month.[5] But this was merely a prelude to a full-scale assault.[6] On the morning of 28 September 1859, a group of men prepared to attack the dam. The sheriff appeared in time to ward off the clash, but by afternoon several pieces of plank were pulled off the dam before French could stop Young and the other intruders. In the course of the struggle, according to Young, Augustus Owen, French's assistant, struck him in the hand with an iron bar. About dark the siege began again as the attackers returned – perhaps as many as fifty – this time with an officer who at Young's insistence arrested French and Owen on complaint of assault and battery. "This of course gave the mob the field for a time," French wrote James Francis, "and they used axes and bars to some purpose, although they did no serious mischief."[7] According to a newspaper account, this last attack proved more successful than the earlier one, "but by the assistance of the 'big boys' of this village, the crowd was summarily dispersed, without much regard to ceremony – some of whom were not handled very lightly."[8]

It was not an easy autumn for New Hampshire's waterpower companies. About a month later, a watchman employed by the Amoskeag Company at Manchester, New Hampshire, spotted a broom handle jutting out from the company's dam across the Merrimack. The watchman carefully drew out the broom handle to discover a flask containing ten pounds of gunpowder connected to fifty feet of fuse. As one report concluded, "It is supposed to have been put there the previous night and was waiting for the right time to come for firing, and thereby blow up the dam and remove the obstructions in the river."[9]

Such events were symptomatic, though not exactly typical, of the conflict raised by the Boston Associates' efforts to control New Hampshire's waters. The flooding of land caused by the construction of large dams became the outstanding focus of such conflict. In addition, the Lake Company caused trouble for others: Mill owners saw water diverted for factories in another state, ferryboat owners found the water in Lake Win-

---

[5] Ibid., 3 Sept. 1859.
[6] The following construction of events is from ibid., 28, 29 Sept. 1859; "Trouble at Lake Village," *Winnipisaukee Gazette*, 1 Oct. 1859; Letter to the Editor, ibid., 8 Oct. 1859.
[7] J. French to J. Francis, 29 Sept. 1859.
[8] "Trouble at Lake Village," *Winnipisaukee Gazette*, 1 Oct. 1859.
[9] "Attempt to Blow Up," *Laconia Democrat*, 4 Nov. 1859.

WOODBURY UNIVERSITY LIBRARY
7500 GLENOAKS BOULEVARD
BURBANK, CA 91510

nipesaukee drawn down too low to navigate safely, and lumbermen were stopped from rafting logs over the company's dams.

The present chapter will explore such conflict, focusing on the 1859 attempt to destroy the Lake Company's dam. This confrontation reflects the tension accompanying industrial change, a struggle over who would control the natural world and to what ends. To examine this event is to sense both the sweeping ambitions of the Boston Associates in their dealings with nature and the determined opposition that emerged to their plans. The conflict over wages and hours, over the enforced discipline of factory life, is a familiar topic to students of nineteenth-century industrialization.[10] I offer this episode as an important parallel to such conflict.

The waters of Lake Winnipesaukee encompass close to 45,000 acres, a prodigious expanse of water that has attracted much attention.[11] When Timothy Dwight, the president of Yale College, visited here in 1812, he looked out across the lake and pictured "an immense field of glass, silvered by the luster which floated on its surface."[12] In others, the lake inspired more practical-minded thoughts. When Nathan Hale, the Boston journalist and author (not to be confused with his uncle who was executed by the British during the American Revolution), took in the view some years later, his mind turned over with the water's potential for production. His was a calculating eye, envisioning the banks of the lake teeming with villagers, "while a thousand mill streams will be drawn off to work the machinery of numerous manufactories."[13]

The lake is fed by several neighboring streams and smaller bodies of water. Lake Wentworth is the largest of these, an oval-shaped patch of over three thousand acres of water. Five islands lie within the lake's fourteen miles of shoreline. About the

---

[10] Indeed, the same year as the attack on the dam, workers at Lowell turned out to demand increased pay. See Thomas Dublin, *Women at Work: The Transformation of Work and Community in Lowell, Massachusetts, 1826–1860* (New York, 1979), 203.

[11] This figure and the ones that follow for the area of New Hampshire's lakes are from Charles J. Swasey and Donald A. Wilson, *New Hampshire Fishing Maps* (Freeport, Me., 1986).

[12] Timothy Dwight, *Travels in New England and New York*, ed. Barbara Miller Solomon, 4 vols. (Cambridge, Mass., 1969), 4:101.

[13] Nathan Hale, *Notes Made During an Excursion to the Highlands of New-Hampshire and Lake Winnipiseogee* (Andover, Mass., 1833), 38.

islands the water stirs ever so slightly as it escapes through a set of smaller bays before reaching the southeastern corner of Lake Winnipesaukee. The waters of Lake Winnipesaukee, Lake Wentworth, and the other main tributary ponds spread out in a vast aqueous sheet across New Hampshire's heartland. But the water here does not stay put.

There is only one way for the water to move out of Lake Winnipesaukee and that is through an outlet at its western end. From this point, the water flows through a series of bays. The water journeys through an elongated lake called Paugus Bay before pausing briefly in the diminutive Opechee Bay and then emptying its full force into Lake Winnisquam. With over 4,200 acres of water, Lake Winnisquam is New Hampshire's third largest freshwater lake. Its plentiful waters gather at its outlet before bulging into another much smaller body of water. Eventually, the water moves forth into the Winnipesaukee River, which joins the Pemigewasset River in Franklin, New Hampshire, to form the Merrimack.

Lake Winnipesaukee and the neighboring bays comprise more than fifty thousand acres of surface water. These waters are the lifeblood of the Merrimack River, an enormous hydrous hinterland that imparts a remarkable uniformity and steadiness to the river's streamflow. And in 1845 the Boston Associates made up their minds to own them.

In that year, there were thirty-one mills in Lowell, and water set in motion over 225,000 spindles and more than 6,300 looms.[14] The number of leased mill powers had grown significantly during the first half of the 1840s. By 1840, the PLC leased sixty-five mill powers. Three years later that number had increased to ninety-one, a gain of 40 percent.[15] Water had become an ever more precious commodity, and concern over it heightened, especially with the founding of the Essex Company. In 1845, anticipating the needs of the new textile city along the lower Merrimack River, Samuel Lawrence, Abbott's brother, turned his attention to the major New Hampshire lakes feeding the Merrimack: Lake Winnipesaukee and the bays and river leading out of it, the Squam lakes, and Newfound Lake. "After the work on the dam at Lawrence had been

[14] Patrick M. Malone, *Canals and Industry: Engineering in Lowell, 1821–1880* (Lowell, 1983), 9.
[15] Louis C. Hunter, *Waterpower in the Century of the Steam Engine,* A History of Industrial Power in the United States, 1780–1930, vol. 1 (Charlottesville, Va., 1979), 268.

commenced," wrote Samuel Lawrence, "I became alarmed lest the control of those grand reservoirs should be in the hands of parties not in harmony with the mill-owners on the main stream."[16]

The lack of water during the early 1840s caused a great deal of apprehension at Lowell. Patrick Tracy Jackson, the PLC's agent at this time, found that several of the mills were running considerably more machinery than before. As a result, the mills "cannot keep up their speed even when water is rendered according to our contracts with them . . . [and] when in a dry time it falls below level, they draw so hard upon the Pawtucket Canal as to cause a great descent in it."[17] Before 1846, according to one report, roughly five feet of head (distance the water falls to the waterwheel) was lost just a few hours after the start up of the Lowell mills, requiring a lower speed for the remainder of the day.[18]

When the Merrimack valley waterpower companies leased water, they did so on the basis of permanent mill powers – the whole capacity of the river during dry times. The number of permanent mill powers was the volume of water always to be counted on, even under the driest conditions.[19] But as Jackson and others realized, the demand for water had surpassed the ability of the river to provide, a problem particularly troubling during the drier summer months from July through September. The solution was more adequate storage. A supply of stored water served the needs of production by creating a steady, reliable flow. In general there were two ways of storing water. Either it could be held directly behind the dam or in lakes or reservoirs upstream from the factory.[20] Until almost midcentury, the Merrimack valley mill towns stored water using the former approach. But at Lowell, the ambitions for large-scale textile manufacturing were outstripping the capacity of the mill pond behind the dam at Pawtucket Falls.

Concern that the river would not live up to the expectations of production spread throughout this industrial community.

16 "Three Letters of Samuel Lawrence, Esq.," *ORHA* 1(1879): 289.
17 P. T. Jackson to Directors, 13 Sept. 1839, vol. A-1, file 7, PLC Papers.
18 "Memorandum about Power at the Appleton & Hamilton Mills," 20 Dec. 1883, vol. A-17, file 83, PLC Papers.
19 J. Francis to B. Saunders, 21 Oct. 1864, vol. DB-3:567, PLC Papers.
20 Robert B. Gordon, "Hydrological Science and the Development of Waterpower for Manufacturing," *Technology and Culture* 26 (1985): 206.

This fear prompted an organizational change in 1845. In that year, the textile companies at Lowell ended the independence of the PLC by taking over its stock. "These companies," wrote James Francis in reference to the move, "fearing the use of the water power might become too much extended, secured the control of the whole remaining water power."[21] The action signified, in a sense, a subtle anxiety over the limits of nature, alarm that production might exceed the river's ability to provide water.

The proximate reason for such concerns is clear: the expansion of productive capacity. More mills meant more spindles and looms to move by using water. But there may have been other factors tending to increase the demand for waterpower. "The general experience at Lowell," in Francis's expert opinion, "has been that without any material increase in the amount of machinery in a Mill, yet in the course of time a considerable increase of power is required."[22] Francis attributed the problem to a change in the machinery used and increased operating speeds. But more subtle processes also worked to expand the need for water. The floors of the mills were built of wood, and over the years the immense weight of the machines caused them to sink. The uneven surface in turn increased the chances for the bearings to bind, a development creating the need for more power.[23]

With the need for water on the rise, the Waltham–Lowell factories placed more pressure on the entire Merrimack watershed. Watersheds are complicated systems, and their available flow is always limited by a number of physical factors. Precipitation is perhaps the most important. In general, nature was reliable in this respect, although some variation existed from year to year. As measured at Lowell between 1834 and 1846 – the latter date being roughly when the Boston Associates began their purchases in New Hampshire – the annual average rainfall amounted to a little less than thirty-six inches; the period from 1847 to 1859, in contrast, averaged considerably more, slightly over forty-four inches.[24] But a point measurement is not representative of areas of more than a few square miles and

---

[21] J. Francis to B. Saunders, 21 Oct. 1864.
[22] J. Francis to H. Hall, 21 Feb. 1853, vol. A-18, file 89, PLC Papers.
[23] J. Francis to I. Hinckley, 21 Feb. 1859, vol. DA-5, PLC Papers.
[24] See rainfall figures at Lowell, *Lowell Daily Courier*, 1 Feb. 1886, a copy of which can be found in vol. ND-1, PLC Papers.

is of limited use in determining the quantity of streamflow in the nineteenth-century Merrimack.[25] Unfortunately, precipitation figures for the Merrimack basin in the period before mid-century are scarce. Unless more data are uncovered, it seems fair to assume that dry conditions alone did not inspire the Boston Associates on their quest for the New Hampshire lakes.

Apart from precipitation, deforestation of a watershed can also affect streamflow. By the nineteenth century, New Englanders had been chopping away at the wooded landscape, peeling it back and converting forests to settlements for some two hundred years. In the Merrimack River basin, deforestation reached a peak between 1860 and 1880 before beginning to recover. Unquestionably, significant quantities of forest had been destroyed there by much earlier in the century. Removing substantial numbers of trees from a watershed, it has been shown, increases water yields. But the effects of such behavior on a large stream like the Merrimack are difficult to assess. One study of the Merrimack valley found no change in water yields despite the deforestation of the basin between 1850 and 1880. Still, one can assume that if deforestation had any effect – and it is hard to believe that it had none – it was to increase the available streamflow.[26]

Environmental factors alone were not responsible for the perceived lack of water at Lowell. Overall, there can be little question that the New England environment, with its steady precipitation and generous water resources, remained congenial to the interests of the region's water-driven factories. The Boston Associates might well have taken solace in this fact had they not had to deal with a man-made factor: their own increasing thirst for more and more water. "Wants," the anthropologist Marshall Sahlins writes, "may be 'easily satisfied'

---

25 On measuring rainfall, see Thomas Dunne and Luna B. Leopold, *Water in Environmental Planning* (New York, 1978), 67.

26 For an introduction to the forest history of New England, see Roland M. Harper, "Changes in the Forest Area of New England in Three Centuries," *Journal of Forestry* 16 (1918): 442–52. Deforestation in the Merrimack valley during the nineteenth century is described in U.S. Congress, "The Influence of Forests on Stream Flow in the Merrimac River Basin, New Hampshire and Massachusetts," prepared by Edward Burr, 62d Cong., 1st sess., 1910, H. Doc. 9, 9. Experiments during the 1920s showed that removing trees from a watershed increased streamflow. See W. G. Hoyt and H. C. Troxell, "Forests and Stream Flow," *Transactions of the American Society of Civil Engineers* 99 (1934): 1–30, and the discussion pp. 31–111; also see Eleanor C. J. Horwitz, ed., *Clearcutting: A View From the Top* (Washington, D.C., 1974).

either by producing much or desiring little."[27] The Boston Associates chose to produce much.

To deal with the concerns over water, Jackson made several suggestions in 1839. He recommended that a substantial leak in the Pawtucket Dam be repaired; he accepted Francis's advice that water be held back during the night to increase the daytime supply; and he proposed the construction of the Northern Canal to provide power for the Lawrence, Suffolk, and Tremont mills. There were originally two plans for the Northern Canal, the directors choosing to build the more ambitious design. The decision for the larger version was based on a predicted flow of 3,800 cfs at a dry time and over 4,000 cfs with a full river. The figures themselves rested on two assumptions. First, that water would be stored overnight as Francis suggested and, second, that control over the New Hampshire lakes feeding the Merrimack would be secured.[28]

In the fall of 1845, Abbott Lawrence, who would soon have a burgeoning textile city named for him, set off to gain control over the waters of Lake Winnipesaukee. His first move involved the purchase of the Lake Company. That company had been incorporated in 1831, with a capital stock of $100,000 to run two textile mills at Folsom's Falls in Lake Village. Lawrence bought the Lake Company in 1845 for $60,000, giving him control of 250 acres of land, factory buildings, and water rights at Folsom's Falls.[29] This marked the start of the Boston Associates' drive to take command of New Hampshire's impressive water resources. Their venture aimed to establish three separate water storage systems to benefit the mills at Lowell and eventually at Lawrence. To do this they needed to make some carefully planned purchases of land and water rights in the region.

27 Marshall Sahlins, *Stone Age Economics: Production, Exchange, and Politics in Small Tribal Societies* (Chicago, 1972), 1–2.
28 P. T. Jackson to Directors, 13 Sept. 1839. This letter mentions the leak in the dam and is an early proposal for a new canal. For the formal recommendation on the plan for the Northern Canal, see P. T. Jackson, J. Baldwin, and C. S. Storrow to J. A. Lowell, E. Chadwick, J. T. Stevenson, and H. Hall, Directors' Records, 10 Apr. 1846, vol. 3, PLC Papers.
29 Lake Co. to A. Lawrence, 21 Oct. 1845, bk. 7:561–4; D. Pingree to A. Lawrence, 21 Oct. 1845, bk. 7:565–6; D. Pingree to A. Lawrence, 21 Oct. 1845, bk. 7:567–8, BCRD. Pingree was the Lake Company's agent at the time.

Two other people, Nathan Crosby and John Nesmith, helped to bring this water into line with the needs of the mills along the lower Merrimack. Both men were born in New Hampshire and had ties to the Boston Associates.[30] Crosby, a neighbor and friend of Samuel Lawrence, evidently was well informed of New Hampshire's water resources. Nesmith was a promoter of the city of Lawrence. In 1846, they assembled a vast array of land and water rights along the bays leading out of Lake Winnipesaukee, and at other points in New Hampshire, thereby securing control of the water as it flowed to the Merrimack River. Their strategic purchases, later transferred to the Lake Company, were calculated to give it the power to govern the Merrimack's flow to aid the mills downstream in Massachusetts.[31]

Meanwhile, Abbott Lawrence established the necessary organizational structure for the new Lake Company. In 1846, he petitioned the Massachusetts legislature to allow the Essex Company to acquire and hold property in the state of New Hampshire. The Essex Company, he explained, was interested "in such improvements as shall cause a steady & continuous flow of water throughout the year in" the Merrimack River. Lawrence was not entirely candid in his request. He stated the company's intention to improve *both* the navigation and the waterpower (written in that order) of the Merrimack. This could only be done, he continued, by expenditures in New Hampshire.[32] In March 1846, the governor signed his request, allowing the company to hold stock in the Lake Company.[33]

In addition, the New Hampshire legislature needed to amend the Lake Company's original charter (granted in 1831) so more capital stock could be issued. This task fell to Charles Storrow. He enlisted the help of a lawyer named James Bell, who in 1846 was elected to the New Hampshire legislature.[34] That same year, Bell assured Storrow that a bill increasing the

---

[30] For biographical information on Crosby and Nesmith, see Charles Cowley, *Illustrated History of Lowell*, rev. ed. (Lowell, 1868), 133, 201–2.

[31] Crosby and Nesmith made many purchases. For some of Crosby's more important transactions, see bks. 7:493, 8:450–5, 467, 469, 474, 482, BCRD. Nesmith's purchases are recorded in bk. 9:230, BCRD. Also see Nesmith to Lake Co., 8 Dec. 1846, bk. 83:201, Merrimack County Registry of Deeds, Concord, N.H.

[32] "Petition to the General Court," 4 Feb. 1846, item 2, EC Papers.

[33] Act of 18 Mar. 1846, ch. 119, [1846] Mass. Acts & Resolves 81.

[34] For a brief biography of James Bell, see Charles H. Bell, *The Bench and Bar of New Hampshire* (Boston, 1894), 190–2.

capital stock of the Lake Company to $1 million would soon be approved. Not long after this happened on 10 July 1846, Bell decided to change jobs.[35] From 1846 to 1855, Bell served as the agent of the Lake Company, playing a vital part in this elaborate plan to divert the waters of New Hampshire into the hands of the Waltham–Lowell mills.[36]

By the summer of 1846, the Lake Company, transformed from a limited local venture into a vastly more ambitious enterprise, began in earnest to take command of a sizable share of New Hampshire's surface water. The board of directors elected by the fall included Samuel and Abbott Lawrence, John A. Lowell, Charles S. Storrow, and Ebenezer Chadwick – a textile investor and a powerful figure in Boston financial circles.[37] When the effort to control the lakes began, the Amoskeag Company at Manchester offered to purchase a one-fifth share. Abbott Lawrence and several others countered by suggesting the Amoskeag Company take a full third share, an offer the Amoskeag directors rejected. It was much later in the century before the Amoskeag Company again considered a stake in the project.[38]

The Lake Company next began to put together the necessary infrastructure for water control. As early as the fall of 1846, James Bell had workers deepening and widening the channel at the outlet of Lake Winnipesaukee. The channel here was lowered several times between 1846 and 1849, resulting in an added depth of four to six feet.[39] This change allowed the Lake Company to more easily draw down the waters of the lake. In

---

35 J. Bell to C. Storrow, 20 June 1846, item 14, EC Papers; also see Act of 10 July 1846, ch. 437, [1842–8] N.H. Laws 436.

36 J. Bell to C. Storrow, 30 July 1846, item 14, EC Papers; also see Bell, *Bench and Bar*, 190–1.

37 Information on the board of directors is in *Cole v. Lake Co.*, 54 N.H. 242, 252 (1874). The facts in the case include a vote of the Lake Co. in 1846 listing the names of the five directors.

38 Directors' Records, 8 Oct. 1845, vol. A-2, AMC Papers, BL. Samuel Lawrence wrote a mysterious note to Charles Storrow regarding the AMC's participation in the plans for the New Hampshire lakes, which reads, in part: "Go to the death against allowing Amoskeag to participate, this grows on me wonderfully & I see cause constantly why it is so." See S. Lawrence to C. Storrow, 9 Oct. 1845, item 204, EC Papers. Late in the 1870s, renewed efforts were made to involve the AMC in the Lake Company venture. See Chapter 8.

39 J. Bell to J. Francis, 16 Oct. 1846, vol. A-37, file 204, PLC Papers. On the repeated deepening of the channel, see the depositions taken in the following legal document in which the excavations are discussed. *Town of Gilford v. W. L. C. & W. Manufacturing Co.* (1871), Deposition, item 204, p. 6, EC Papers.

1851, the company completed a new dam about 250 feet in length at Folsom's Falls to replace the old boulder and rubble dam built in 1829. Constructed of stone and almost exactly the height of the old dam (approximately ten feet), the new one was tighter, and flooded more nearby land.[40] The dam served as the primary structure for controlling the water in Paugus Bay and Lake Winnipesaukee.

A system of gates and sluices at the dam was used to regulate the flow of water from Lake Winnipesaukee into the bays below. When engineers at Lowell or Lawrence called for water, the company raised the gates of the Pearson Dam at the outlet of Lake Winnisquam, allowing the water to pass downstream to the Merrimack River. Simultaneously, it opened the gates at the Lake Village Dam to replace the water that had been released. This procedure shortened the wait for water in Massachusetts by eliminating delays in filling the intermediate bays.[41]

For its plan to work, the Lake Company needed control over the dams below the one in Lake Village. Dams crossed the Winnipesaukee River in three main areas. By the early 1850s, the company had negotiated with local milldam owners for the right to control the flow along the thirteen-mile stretch from the outlet of Lake Winnipesaukee to the Merrimack River.[42]

The Lake Company's Winnipesaukee water storage system seemed complete by 1851. But an important addition to it was made three years later. In the summer of 1854, James Bell purchased mills, land, and water rights at the outlet of Lake Wentworth.[43] The lake's water flowed briefly through a pond

[40] A description of the Lake Company's dam and its history from 1829 to 1851 is found in *Gilford v. Winnipiseogee Lake Co.*, 52 N.H. 262–3 (1872). Also see *Winnipissiogee Lake Cotton and Woolen Man'g Co. v. John L. Perley* (1865), Depositions, pp. 42–4, N.H. Reservoirs Pamphlet Box, Special Collections, University of Lowell, for Stephen C. Lyford's account of the history of the dam. Lyford became interested in the waterpower at Lake Village and contributed to the construction of the 1829 dam. For a description of the dam in 1849, see J. Francis and Captain Bigelow to Directors of Lake Co., 11 Dec. 1849, vol. DA-2C, PLC Papers.

[41] George F. Swain, "Water-Power of Eastern New England," in U.S. Department of Interior, Census Office, *Reports on the Water-Power of the United States*, vols. 16 and 17 (Washington, D.C., 1885–7), 16:52 (hereafter cited as Swain, "Water-Power of Eastern New England").

[42] For $11,000, the owners of water rights at the Avery Dam in Laconia, N.H., promised to manage the water there as the Lake Company wished. See Lake Co. and Belknap Manufacturing Company, Gilford Manufacturing and Mechanic Company, and F. Boynton, 14 Oct. 1852, bk. 20:89, BCRD. Also see Theodore L. Steinberg, "Nature Incorporated: The Waltham–Lowell Mills and the Waters of New England" (Ph.D. diss., Brandeis University, 1989), 211–12.

[43] J. Bell to J. Francis, 28 July 1854, vol. A-37, file 202; J. Bell to J. Francis, 2 Sept. 1854, vol. A-37, file 204, PLC Papers.

before passing into Lake Winnipesaukee. Plans were later made for excavating the channel between Lake Wentworth and the pond, and two years later, a dam at the outlet of the pond was completed.[44] "The purchase of this pond and the control of the water," wrote Josiah French, Bell's successor as the company's agent, "was undoubtedly very important, in view of having something in reserve, to supply the great Lake [Winnipesaukee] in case of a long continued drought."[45] The addition gave the company an extra reserve of over three thousand acres of water.

James Francis often visited New Hampshire to observe the progress of the storage systems. In 1852, he returned from one such visit to write to Henry Hall, the president of the Lake Company. "Mr. Bell," he wrote, "expresses himself strongly on the importance of not relying exclusively as heretofore on Lake Winnipiseogee as a reservoir, considering also that the Mills at Lowell now require a greater supply of water than ever before." Francis agreed that the Lake Company should diversify its system of water storage.[46] During the 1850s, the company's ambitions for water control spread out across the waterscape of New Hampshire.

The Squam lakes lie just a few miles northwest of Lake Winnipesaukee, but their waters take a much different path before reaching the Merrimack River. Settled in below the towering Squam Mountains, Squam Lake consists of over 6,700 acres of water. The water here flows southwest to Little Squam Lake (roughly 400 acres) before coursing two and one-half miles to the Pemigewasset River, twenty-three miles from where the Merrimack has its start.[47] Nathan Crosby and James Bell both bought land in this area for the Lake Company.[48] By 1848–9, a fifteen-foot dam had been constructed at the outlet that raised the water in the two lakes by several feet.[49] Excavations in the channel between the lakes and at the outlet of Little Squam

---

[44] J. Bell to J. Francis, 16 Sept. 1854, vol. A-37, file 204, PLC Papers; J. French to F. B. Crowninshield, 3 Mar. 1859, item 204, EC Papers.

[45] J. French to F. B. Crowninshield, 3 Mar. 1859.

[46] J. Francis to H. Hall, 25 June 1852, vol. DA-3, PLC Papers.

[47] The figures are from Swasey and Wilson, *New Hampshire Fishing Maps*, 61–2; Swain, "Water-Power of Eastern New England," 55.

[48] J. French to J. Thomas Stevenson, 2 Dec. 1862, item 204, EC Papers. Stevenson was treasurer of the Lake Co.

[49] J. French to F. B. Crowninshield, 3 Mar. 1859.

improved the flow of water. The Lake Company by 1854 had a second system for storing water in place.[50]

A final system operated at Newfound Lake about ten miles west of Lake Winnipesaukee. This 4,100-acre expanse of water drains into the Newfound River before flowing south to the Pemigewasset. Although the surface of this lake is substantially smaller than that of Squam Lake (close to 40 percent smaller), the drainage area of Newfound Lake (about ninety-one square miles) is significantly larger. For this reason, Newfound Lake – once the Lake Company built a dam there in 1852 – probably yielded more water than the Squam lakes.[51] Under James Francis's supervision, the dam here was rebuilt in 1858–9. Eight feet in height, the new dam raised the water of the lake by about the same amount. Across the top of the structure, a building designed to keep away all would-be intruders enclosed the machinery for operating the gates. During this time, the channel of the river below the dam was lowered an additional three feet. The company negotiated with dam owners below on the Newfound River for the privilege of storing and releasing the water in the lake as it wished.[52]

By 1859, the Lake Company had its three systems of water storage at Winnipesaukee, Squam, and Newfound lakes in operation. But the storage systems it created were much less efficient than the old mill pond directly behind the dam at Lowell. Because the New Hampshire lakes were far from Lowell and Lawrence (the lakes ranged from about 80 to 100 river miles from Lowell), it took a good deal of time – several days – for the water to pass downstream. Water directly in back of the Pawtucket or Lawrence dams, in contrast, was available immediately. To deal with the distance problem, the Lake Company and its sponsors, the PLC and the Essex Company, had to develop an efficient system of communication. That system took some time to perfect. Before 1858, James Francis simply wrote

---

[50] J. Bell to J. Francis, 2 Oct. 1854, vol. A-37, file 204, PLC Papers.

[51] Dwight Porter, "Water Power of the Merrimac River and of Rivers of Eastern Massachusetts between the Merrimac and the Blackstone" (ms. at MATH, 1899), 57, 60–1.

[52] See J. French to F. B. Crowninshield, 3 Mar. 1859. This letter describes the rebuilding of the dam in 1858–9 and the payment to dam owners below on Newfound River. On the height of the dam and the deepening of the channel, see the facts in *Benj. F. Holden v. Winnepisseogee Lake Cotton & Woolen Mfg. Co.* (1872), vol. 93:435, NHSCN. Many of the nineteenth-century New Hampshire cases that reached the state's highest court have unpublished notes filed in the New Hampshire Supreme Court Library.

to James Bell with his impressions of the river and the prospects for waterpower in the near future.[53] But in the summer of 1858, Francis arranged a more rationalized system with the new agent, Josiah French. Instead of giving a general impression of the need for water at Lowell, Francis devised a numeric system that estimated more exactly the water that would be required. He let each digit from one to twenty-five represent 100 cfs. In Francis's view, 2,500 cfs was needed night and day to run the mills. If French received the number twenty, for example, he could assume that the Lowell mills would be short 500 cfs and make the necessary changes to the water at the lakes.[54]

The system was never perfectly effective, not with the lakes at such a distance from the lower Merrimack River. But control of the New Hampshire lakes unquestionably supplied Lowell and Lawrence with more water than the river naturally would have conferred. Forty-seven more mill powers were divided among the factories at Lowell, bringing the total number to just over 139 in 1853. This was more than 50 percent above the number of mill powers leased in 1845 before the acquisition of the lakes.[55] The newly built Northern Canal – completed in 1848 – delivered some of this newfound water wealth. James Francis believed the increase in waterpower resulted in part from the improvements in the Lowell canal system, but mainly from the New Hampshire purchases.[56] In his opinion, control of the lakes increased the dry season flow (between July and October) by two or three times what it had been.[57] The gain was impressive and so was the Lake Company's substantial holdings of land and water in New Hampshire. In 1859, Francis calculated the area of surface water controlled by the Lake Company to be over 103 square miles – a veritable hydraulic empire.[58]

Some months before the attack on the Lake Village Dam, Josiah French made a prescient remark. Writing to the Lake Company's treasurer, F. B. Crowninshield, he reflected:

> The construction of dams, the widening and deepening of the streams and other improvements of these several Reservoirs, without

[53] For examples see any of the letters from J. Francis to J. Bell written after 1847 in vols. DA-2C, DA-3, DA-4, PLC Papers.
[54] J. Francis to J. French, 30 Aug. 1858, vol. A-17, file 82, PLC Papers.
[55] Memo Regarding New Hampshire Reservoirs, 17 Mar. 1859, vol. A-17, file 82, PLC Papers.
[56] J. Francis to Directors, 10 July 1866, vol. A-17, PLC Papers.
[57] J. Francis to A. Penfield, 7 June 1867, vol. DB-5:200–201, PLC Papers.
[58] Memo Regarding New Hampshire Reservoirs, 17 Mar. 1859.

any other ostensible purpose than the control and use of the water at some distant point, apparently foreign to the interest of of [*sic*] the people on the Lakes or streams is, and will be more and more, looked upon as an infringement upon the interests of their neighborhood, town, and state, if not upon their individual rights, and the feverish state of public feeling upon the subject of water rights, and the fact that a great portion of the large bodies of water in the state are owned by the Lake Company, make it important to guard as much as possible against increasing a feeling that is ready to enlist in any crusade against any and all corporations where there is the least ground.[59]

The events of the following autumn would prove French correct. He properly sensed the rising tide of opposition toward the Lake Company. But then again, it was a tide that had been rising for some years.

Conflict over water was not new to the Boston Associates. When the PLC at Lowell added two feet of granite to the Pawtucket Dam in 1833 (see p. 65), neighboring landowners complained about flooding. By the 1840s, the PLC had raised the dam two feet higher with flashboards, overflowing land upstream in Massachusetts and New Hampshire. But the PLC managed to contain the conflict by making settlements before most landowners took legal action.[60]

The Lake Company venture generated conflict on an entirely different order. At first, most of the tension concerned the flooding of land in New Hampshire. During the 1850s, farmers accused the company of flooding their meadows. From late July into early August, farmers routinely cut hay from their land on the Winnipesaukee River downstream from the dam at Lake Winnisquam. The trouble developed as the Lake Company released water from the bay to supply the Massachusetts factories, flooding the meadows and preventing local farmers from harvesting the hay. To give farmers a chance to cut their hay, James Bell sometimes used the company's dams to hold back the water along the Winnipesaukee River. To satisfy the Massachusetts mills, he made up the difference by allowing more water to escape from Newfound Lake.[61]

---

59 J. French to F. B. Crowninshield, 28 Feb. 1859, item 204, EC Papers.
60 The following letters discuss the difficulty of settling claims for flooding: K. Boott to D. Abbot, 25 June 1836; K. Boott to L. Drake, 9 July 1836; W. Boott to C. Blood, Jr., 22 June 1841; W. Boott to D. Abbot, 29 Sept. 1842; W. Boott to D. Abbot, 29 Nov. 1842; W. Boott to J. W. Parker, 29 Nov. 1842; W. Boott to D. Abbot, 9 Dec. 1842; W. Boott to J. M. Parker, 5 June 1843, vol. DA-2C, PLC Papers. Also see "Memorandum about Power," 20 Dec. 1883.
61 J. Bell to J. Francis, 25 July 1853, 10 Aug. 1854, vol. A-37, file 204, PLC Papers.

Resistance increased as the Lake Company further developed the region's lakes, especially as it moved forward after 1852 with plans for Newfound Lake and Lake Wentworth. In July 1855, Bell wrote James Francis of "a number of persons, principally irresponsible young men" who had demolished the company's coffer dam (a temporary structure erected in this case to hold back water while excavations in the streambed took place) at Lake Wentworth.[62] Later that same summer, owners of meadowland on Fowler's River – which emptied into the western side of Newfound Lake – blamed the company for flooding their lands and claimed over $1,200 in damages. When Bell wrote Francis about the matter, he singled out one landowner, in his opinion "a violent and reckless man," who had made repeated threats to destroy the company's dam. Bell took the threats seriously, concluding that "there is reason to apprehend that he may do it."[63]

More intense opposition toward the Lake Company broke out after 1855. In two separate legal cases, the plaintiffs tried to stop the company from proceeding with its designs for water control. John Coe was a wealthy landholder with 150 acres of land bordering Squam Lake.[64] When the Lake Company constructed its dam at the outlet of the lake in 1848–9, it had settled with nineteen landowners. But Coe refused agreement, and in 1857 he brought his case before the Supreme Judicial Court of New Hampshire (SJC) – the state's highest court.[65]

Coe asked the court to enjoin the Lake Company from excavating in the channel between the Squam lakes and below in Squam River. If the company proceeded with its plan, he believed his land, particularly his trees and grass, would be destroyed.[66] The company filed a demurrer in response. This is a legal step taken by a defendant to have a suit dismissed on the grounds that the plaintiffs do not have a right to relief even if the facts alleged are true. In this case, the company argued that Coe had no right to appeal for an injunction because it is an extraordinary remedy, an action of last resort. The court agreed to dismiss Coe's bill the following year, denying the

---

62 Ibid., 23 July 1855, vol. A-37, file 202, PLC Papers.
63 Ibid., 2 Aug. 1855, vol. A-37, file 202, PLC Papers.
64 For a brief biography of Coe, see D. Hamilton Hurd, *History of Merrimack and Belknap Counties, New Hampshire* (Philadelphia, 1885), 728–9.
65 J. French to F. B. Crowninshield, 3 Mar. 1859. Prior to 1855, the court had been called the superior court of judicature.
66 For a copy of Coe's bill for an injunction, see *John Coe v. The Winnipisiogee Lake, Cotton & Woolen Manufacturing Company* (1857), item 204, EC Papers.

equity court's jurisdiction in the case. If the Lake Company did indeed cause damage to Coe's land, the court ruled, he could be compensated by a monetary award. Under the circumstances, an injunction was not necessary.[67]

John Parker brought a similar case against the company. Parker owned part of a water privilege in Meredith Bridge (Laconia), downstream from the Lake Village Dam. The site had been used at times to turn two or three lathes for making bedsteads.[68] In 1846, he offered to sell Abbott Lawrence his share in the privilege for $6,000 – an offer Lawrence apparently declined.[69] In his suit against the company, Parker argued that the Lake Company's dams and excavations damaged his waterpower by diverting the water as it flowed naturally out of Lake Winnipesaukee. Because he was living in New York, Parker brought his suit in federal court. Again, the Lake Company demurred. When the case came to trial in 1859, the court argued that the case lay outside the jurisdiction of a court of equity and dismissed Parker's bill for much the same reasons as in Coe's case.[70]

In Parker's view, the company's Winnipesaukee water storage system damaged his waterpower downstream. But did it? The company's attorney in the case, William H. Y. Hackett, argued otherwise. The Lake Company's dams and excavations along the Winnipesaukee River, according to Hackett, actually improved Parker's waterpower. By storing water and releasing it during dry periods, the company had secured "a comparative uniformity in the flow of water through the year – to the benefit of their mills and of every one interested in any and every water wheel between the Lake [Winnipesaukee] and the ocean."[71] It was not the Lake Company's intention to improve

---

[67] *Coe v. Winnepisiogee Lake Cotton & Woolen Mfg. Co.*, 37 N.H. 254, 266–7 (1858).

[68] J. Bell to Directors, 15 Dec. 1855, vol. A-37, file 202, PLC Papers.

[69] J. A. Parker to A. Lawrence, 7 Feb. 1846, item 14, EC Papers.

[70] *Parker v. Winnipiseogee Lake Cotton & Woolen Mfg. Co.*, 18 F. Cas. 1181 (C.C.D. N.H. 1859) (No. 10,752). The opinion denied the jurisdiction of a court of equity to interfere unless irreparable damage or an interminable amount of litigation might otherwise result. Equity courts would render decisions in situations in which courts of law were unable to administer an adequate remedy, i.e., where money damages would constitute inadequate relief. Parker eventually appealed the case to the U.S. Supreme Court where the lower court's decision was affirmed. See 67 U.S. (2 Black) 545 (1862).

[71] *Argument of William H. Y. Hackett, for the Defendants, in the Suit in Chancery, John A. Parker against the Winnipiseogee Lake Cotton and Woolen Manufacturing Company* (Portsmouth, N.H., 1859), 22.

the waterpower available in New Hampshire. But its actions may well have had this effect. Contrary to Hackett's assertions, however, the company's endeavors probably had less positive impact on the Winnipesaukee River than elsewhere. The plentiful waters of Lake Winnipesaukee impart a uniform flow to the river below. Thus the Lake Company's redesign of the waterscape was not felt as strongly there as, for example, on Newfound River. That river would often run low during dry summers. But through excavations and careful management of the water, the Lake Company made the natural flow more continuous and reliable.[72] In the end, the factories at Lowell and Lawrence benefited, but so did the local mills in the region.

None of this appears to have impressed Parker. According to French, when Parker realized his legal prospects were not terribly bright, he drew up a petition to the New Hampshire legislature.[73] In June 1857, he presented the petition bearing an impressive list of 929 names – people living in Laconia, Meredith, Sanbornton, Gilford, and other towns bordering the state's vast waters. The petition accused the Lake Company of misrepresenting its intentions. The company had been established under its original charter (passed in 1831, but amended in 1846 to allow more capital stock) to manufacture textiles. The petitioners maintained that the company had abandoned this objective in 1846 and "ever since without any legal right or authority so to do continued to devote all their capital and resources to the acquisition and maintenance of control over the waters of the Winnipissiogee Lake and River, and the tributaries of the Pemigewasset."[74]

Their complaints did not end there. The Lake Company, they insisted, had managed the water "to the great detriment and loss to the owners of lands and mills" in the region. They had, in short, been economically wronged by the company. The company's actions, in their opinion, had deprived the region of "a large share of the capital now invested in Lawrence and Lowell." In other words, but for the actions of the Lake Company, large textile towns would have been built along the Win-

---

[72] See Swain, "Water-Power of Eastern New England," 51, 55; Richard W. Musgrove, *History of the Town of Bristol, New Hampshire* (1904; reprint, Somersworth, N.H., 1976), 362.

[73] J. French to F. B. Crowninshield, 28 Feb. 1859.

[74] Petition to the New Hampshire Legislature, 1857, attached to J. French to F. B. Crowninshield, 28 Feb. 1859.

nipesaukee River. The Lake Company's management of the region's water resources evidently had robbed the area of its economic potential.[75] One of the area's most valuable resources was literally drained away to benefit powerful corporations in another state. They asked the legislature to intervene, but their complaints apparently withered. No direct action appears to have been taken.

Why the issue died in the legislature is unclear, but it may have been related to the validity of the petitioners' claims. As already noted, the Lake Company may have improved the area's waterpower – a move that certainly contributed to the local economy. Further, no evidence suggests the Boston Associates ever intended to build major mill towns in the Winnipesaukee region. This part of New Hampshire was probably perceived to be too distant from any major commercial center to warrant serious consideration as a textile city. The claims then of the petitioners were vulnerable. Of course, those who signed the petition may well have believed the statements. Indeed, it is possible that the company's control of Lake Winnipesaukee foreclosed on the eventual development of local industry. Whatever the case, the company had clearly created a great deal of hostility, anger that eventually coalesced in violence. But we must search elsewhere for the sources of antagonism.

Not everyone who lived near the New Hampshire lakes could benefit from the improved waterpower. For the company's plans for water to help, one needed to own property, specifically a mill that could take advantage of better water control. But the group of men who attacked the company's dam owned no mills, indeed most of them owned little if any property at all. Their animosity toward the company then had little to do with whether the company had or had not improved the waterpower of the region. Rather, it seems that their attack on the dam may have in part resulted from economic frustration, of lives caught up in the capitalist transformation of the region – an economic shift that left them behind.

Agriculture dominated the local economy of Meredith, Gilford, and Laconia – where most of the opposition to the company surfaced – during the mid-nineteenth century (Figure 4.1). (Lake Village itself, the site of the 1859 attack on the dam,

[75] Ibid.

Figure 4.1. A view of Meredith, New Hampshire, 1848. (Courtesy of the New Hampshire Historical Society.)

was not an official municipality.)[76] Over 40 percent of the people living in Meredith and Gilford were farmers in 1850 (Table 4.1). At that time farmers depended on dairying for a living, producing cheese and butter for markets in nearby towns and for the more distant seacoast city of Portsmouth, New Hampshire. Farmers also produced significant quantities of wool, grazing sheep in the highlands along the eastern part of Gilford. Most of the wool was probably used locally, particularly in Meredith where there were four woolen mills in 1850.[77]

This kind of agriculture, based on the production of dairy products and wool for market, was labor intensive. Farmers hired laborers to help make butter or clear land for English hay so there would be fodder for their livestock. The farms were self-sufficient neither in labor nor in food and material goods. Rather, they were – and had been for most of the nineteenth

[76] Originally, half the village belonged to Meredith and half to Gilford. When the town of Laconia was formed in 1855, it assumed Meredith's half of the village.
[77] Manuscript Schedules, U.S. Census of Agriculture, Gilford and Meredith, N.H., 1850, 1860; Hurd, *Merrimack and Belknap Counties*, 755; Manuscript Schedules, U.S. Census of Industry, Meredith, N.H., 1850.

Table 4.1. *Occupational structure: Meredith and Gilford, New Hampshire, 1850*

Occupation	Number	Percentage of total work force	Percentage of total wealth
Farming	766	46.17	70.37
Building	130	7.83	4.19
Manufacturing	277	16.69	10.62
Metal	57	3.43	2.10
Mechanic	38	2.29	1.07
Clothing	73	4.40	3.02
Home furnishing	20	1.20	0.83
Food	3	—	—
Jewelry	3	—	—
Printing and art	13	0.78	0.08
Transport trades	22	1.33	1.91
Building material	12	0.72	0.40
Other trades	36	2.17	1.15
Transport	42	2.53	1.40
Commerce	42	2.53	4.69
Professions	56	3.37	4.89
Nonprof. services	29	1.75	0.64
Unskilled and semiskilled labor	310	18.67	2.07
Public service	7	0.42	1.13

Source: Manuscript Schedules, U.S. Census of Population, 1850. The data are for all adult males eighteen and over. Wealth is measured by the reported value of real estate owned. The form of this table and Table 4.2 is a variation of that used in Michael B. Katz, "Occupational Classification in History," *Journal of Interdisciplinary History* 3 (1972): 83.

century – interdependent affairs, relying on markets for the sale of their commodities. The better off among them loaded their ox carts with butter and cheese and set off for Portsmouth to return with sugar, salt, and alcohol. At one point, many farmers produced clothing and other items themselves. But by the mid-nineteenth century, time so spent was clearly on the decline. Between 1850 and 1860, the value of homemade manufactures dropped over 80 percent in Gilford to virtually nothing, suggesting greater reliance on store-bought goods.[78]

The lumber industry also formed an important component

[78] Hurd, *Merrimack and Belknap Counties*, 755; *One Hundred Anniversary of the Incorporation of Gilford, New Hampshire: Proceedings and Addresses at the Centennial Celebration, June 17, 1912* (Lakeport, N.H., 1918), 55; Manuscript Schedules, U.S. Census of Agriculture, Gilford, N.H., 1850, 1860.

of the local economy. During the winter, logs were cut and hauled away to be sawed into planks and boards during the spring. In the towns of Gilford and Meredith, nine sawmills operated at least eighteen saws in 1850. The industry boomed after fire swept through the Belknap Mountains in the mid-1850s, with lumbermen salvaging the best of what remained.[79] At times, disputes arose as rafts of timber would need to be run over the Lake Village Dam. Ironically, James Bell in 1847 refused to allow lumber to pass over the dam, but found out later, much to his embarrassment, that the Essex Company had purchased the timber to help build its textile city at Lawrence. In explaining his actions to Charles Storrow, Bell mentioned "that the establishment of a right of running logs without permission . . . would be found to interfere very materially with the object of retaining a full reservoir of water."[80]

To lumbermen, water served as an artery of transportation, much as it did for others living around the perimeter of Lake Winnipesaukee. Before the 1830s, sailboats and horseboats – a craft powered by horses on a treadmill mechanism – were the primary means of navigation. The boats were used mostly for trade, transporting boards, flour, fish, and molasses to the towns along the rim of the lake. After the 1830s, steamboats made their appearance, increasing in number during the following decades[81] (Figure 4.2). When the Lake Company drew down the waters of Lake Winnipesaukee to supply factories in Massachusetts, it hindered navigation. At the time of the Lake Village riot, for example, the Dover Steamboat Company claimed damages from the company, presumably for creating dangerous low-water conditions.[82]

The most important demand for water came from the industry in the region. In 1860, the towns of Meredith, Laconia, and Gilford had a combined total of thirty-five waterpowered

79 *One Hundred Anniversary of the Incorporation of Gilford,* 58–9; Manuscript Schedules, U.S. Census of Industry, Gilford and Meredith, N.H., 1850.
80 J. Bell to C. Storrow, 30 June 1847, item 14, EC Papers. In 1860, when the Lake Co. had been storing water in Newfound Lake in anticipation of the summer season, a logger threatened to run 130,000 logs through the company's dam without consent. Since the Lake Co. could not risk the loss of stored water, French offered to allow only 10,000 logs through for a sum of twenty-five cents per thousand. See J. French to J. Francis, 7 June 1860, vol. A-38, file 205, PLC Papers.
81 On the use of Lake Winnipesaukee for transportation, see Bruce D. Heald, *Steamboats in Motion* (Meredith, N.H., 1984), 33–5, 39–45, 82–4; Paul H. Blaisdell, *Three Centuries on Winnipesaukee* (Concord, N.H., 1936), 18–20, 24–5, 27–8, 30; Hurd, *Merrimack and Belknap Counties,* 817–19.
82 J. French to F. B. Crowninshield, 3 Mar. 1859. In the 1860s and 1870s, periodic conflict surfaced between navigators and the Lake Co. See B. R. Pipes to Lake Co., 5

Figure 4.2. A steamboat plying the waters of Lake Winnipesaukee near the town of Centre Harbor. (Courtesy of the New Hampshire Historical Society.)

mills. Most of the industry clustered on the Winnipesaukee River in Lake Village and Meredith Bridge, the major business districts in the area.[83] Nine textile and hosiery factories in Gilford and Laconia created the greatest demand for waterpower, although the wood products industry with six sawmills and two shingle mills used large amounts.[84] The Lake Company, for its part, leased waterpower in Lake Village and Meredith Bridge. It rented two textile mills and provided waterpower in Lake Village to a machine shop and foundry. It also leased waterpower to a sizable freight car factory in Meredith Bridge.[85]

The Lake Company tried to portray itself as a contributor to the local manufacturing economy. To soothe tensions and cover its true purposes, it even built a few small mills. In Lake Village

Oct. 1865, and J. French to J. Francis, 5 Oct. 1865, vol. A-39, file 215; J. P. Hutchinson to J. Francis, 12 Dec. 1876, vol. A-45, file 257, PLC Papers.

[83] Like Lake Village, Meredith Bridge was not an official municipal entity. It eventually became part of Laconia.

[84] Manuscript Schedules, U.S. Census of Industry, Gilford, Laconia, and Meredith, N.H., 1860.

[85] Evidence that the Lake Co. leased water is found in legal documents. In the 1870s, the company was sued by one of its lessees, Benjamin J. Cole. The facts in the case explain that the Lake Co.'s directors voted to let James Bell lease unoccupied waterpower sites for periods not exceeding ten years. See *Cole* v. *Lake Co.*, 54 N.H. at 252.

and Meredith Bridge, it had factories and waterpower to rent. Elsewhere, it had to construct new mills to appear devoted to the local community. In 1858, the company erected a sawmill and a gristmill in Bristol, New Hampshire, at the outlet of Newfound Lake. French clearly stated the underlying logic behind the move:

> The object in erecting these mills was not that they were to make a good return for the money expended in their erection, but to do away in some degree the clamor about the Lake Company controlling all the large bodies of water in New Hampshire, for the especial benefit of mills in Massachusetts.[86]

Local residents probably saw through such thinly disguised motives. In any event, the company's scheming probably had little positive impact on the shape of the regional economy. If the towns of Meredith, Gilford, and Laconia are any indication, that economy was in flux by the mid-nineteenth century. As Tables 4.1 and 4.2 indicate, the local economy of the Meredith–Gilford area diversified in the ten years from 1850 to 1860. A smaller percentage of the work force farmed the land, although it was still the predominant occupation. There also occurred a rise of close to 7 percent in the manufacturing work force, with much of the increase accounted for by the clothing industry. Smaller gains were made in the still fledgling areas of transportation and commerce. Although no revolution in the mode of production occurred, the economy was shifting gradually toward a greater emphasis on manufacturing and business.

Another important aspect of the local economy was its wealth distribution, which can be established by comparing the percentage of property owned with the percentage of the work force for each occupation. Farmers possessed the bulk of the property wealth in both 1850 and 1860, but their share of such wealth declined relative to their representation in the work force. Note also that people in manufacturing, although they made up 16 and 23 percent of the work force in 1850 and 1860, respectively, possessed barely more than 10 and 16 percent of the wealth in these years. Professionals secured a sizable share of the total property wealth by 1860 relative to their numbers. But over the ten-year period, laborers, who comprised a significantly smaller share of the overall working population, held a mere 2 percent of the wealth. The data demonstrate a number of instances of wealth inequality, evidence of an economy that did not reward its participants equally.

[86] J. French to J. T. Stevenson, 28 Dec. 1861, item 204, EC Papers.

Table 4.2. *Occupational structure: Meredith, Gilford, and Laconia, New Hampshire, 1860*

Occupation	Number	Percentage of total work force	Percentage of total wealth
Farming	667	42.67	57.81
Building	113	7.23	5.33
Manufacturing	364	23.29	16.29
Metal	71	4.54	2.62
Mechanic	65	4.16	2.85
Clothing	142	9.09	5.37
Home furnishing	4	—	—
Food	14	0.90	0.54
Jewelry	1	—	—
Printing and art	10	0.64	0.29
Transport trades	29	1.86	1.81
Building material	11	0.70	0.35
Other trades	17	1.09	1.14
Transport	68	4.35	2.31
Commerce	72	4.61	5.91
Professions	62	3.97	7.98
Nonprof. services	26	1.66	0.63
Unskilled and semiskilled labor	181	11.58	2.16
Public service	10	0.64	1.58

Source: Manuscript Schedules, U.S. Census of Population, 1860. The data are for all adult males eighteen and over. Wealth is measured by the reported value of real estate owned.

Census information for 1860 is available for nine of the men who attacked the dam the year before.[87] Seven of the nine were living in either Gilford or Laconia, one lived in Northfield, and the other in Concord, New Hampshire. Between them they owned $9,000 worth of property. Much of the land belonged to the wealthiest among them, a farmer named Thomas Plumer ($3,500). James Worster valued his real property at $2,500, but most of his land was probably leased or heavily mortgaged.[88] Of the others, one owned land valued at $1,200, three owned

[87] Manuscript Schedules, U.S. Census of Population, Concord, Gilford, Laconia, and Northfield, N.H., 1860.
[88] See J. Bamford to J. Worster, 18 Oct. 1849, bk. 14:506; J. Worster with E. and W. Barker, 23 June 1851, bk. 17:489; G. W. Young to J. Worster, 8 Oct. 1851, bk. 18:129; E. and W. Barker to J. Worster, 10 Oct. 1851, bk. 18:130; J. Worster to E. and W. Barker, 3 Feb. 1852, bk. 18:540; J. Worster to S. M. Worster, 19 Apr. 1853, bk. 21:91; J. Worster to S. M. Worster, 19 Apr. 1853, bk. 21:92; I. M. Towl to J. Worster, 22 July 1856, bk. 27:313, BCRD.

$500, one owned $300, and two owned no property at all. Thus most of them owned little or no land.

In addition, many of the rioters appear near the bottom of the region's occupational structure. There were two farmers among them, an occupation that in 1860 still accounted for over 50 percent of the total wealth in the area (although the percentage had declined since 1850). The others had jobs in sectors in which their relative percentage of the total work force outstripped the share of total wealth they owned. They worked in manufacturing, transport, and general labor – all occupations with less property than their numbers suggest they should have held.

The 1859 riot was not simply a case of this society's economically dispossessed rising up in arms. A more complicated scenario is suggested by an examination of the motives of those who attacked the dam. But first it is worth asking why sustained opposition to the Lake Company did not emerge earlier. By 1855, the company had been involved in the area for roughly ten years. Why the delay? A clue is found by considering how the Lake Company gained control of land and water rights. It seems doubtful that Crosby and Nesmith explained their intentions when they made the necessary purchases for the Lake Company. It was in the company's interests to keep secret its ultimate objective lest landholders become suspicious and demand higher prices for their property. Moreover, since the purchases were quite disparate – a patch of land in Sanbornton, a piece in Bristol, and so forth – the company's motives were hard to figure out. The natural configuration of the water resources required the company to purchase land and water at points quite distant from each other. And since knowledge and information traveled more slowly than in later decades, it probably took a while for people to realize the company's true ambitions.[89]

By the 1850s, local residents began to make the connections. The Lake Company made more purchases and further refined

[89] "It is estimated," writes Allan Pred, "that within half an hour 68 percent of the American population was aware of President John F. Kennedy's assassination. In sharp contrast, a time lag of seven days occurred between George Washington's death on December 14, 1799, in Alexandria, Va., and publication of that news in New York City." Society in the nineteenth-century Winnipesaukee region lay somewhere in between these two points. See Allan R. Pred, *Urban Growth and the Circulation of Information: The United States System of Cities, 1790–1840* (Cambridge, Mass., 1973), 12–13.

its waterpower infrastructure, digging streambeds deeper and rebuilding dams. The work brought it into contact with larger segments of the local population. With the company's intentions becoming more visible, some began to band together to pursue their interests collectively. John Coe is said to have held meetings near the Squam lakes to unite local landowners in their negotiations with the company over flooding.[90]

In the meantime, the Lake Company's management changed. James Bell, a New Hampshire-born attorney who practiced for six years in the nearby town of Gilmanton, was replaced as agent by Josiah French, a public figure who hailed from Massachusetts. A local historian recalled that Bell "possessed wonderful tact" in his dealings with landholders who found their property flooded by the Lake Company.[91] Perhaps even more important, Bell remained committed to the local community. He lived in Gilford and served as moderator of town meetings from 1849 until 1855 when he left to serve in the United States Senate.[92] An able lawyer dedicated to local affairs, Bell seemed the perfect choice to handle the delicate matters arising over the Lake Company's water control projects.

When James Bell left in 1855, the Lake Company began looking for a new agent. Two years later they found their man. Josiah Bowers French was fifty-seven when he began fourteen years of service with the Lake Company. The son of a Massachusetts farmer, French, by age twenty-four, had moved to Lowell and had been appointed the deputy sheriff for Middlesex County. From this time until 1849 when he was elected mayor of Lowell, French cultivated a long career of public service: coroner, tax collector, assessor, legislator, and fire chief.[93] A veritable factotum, he was a man brimming with ambition who clearly had a great deal of administrative experience, perhaps even a fair amount of savvy. But he was also an outsider, the representative of a company whose heart was with factories in another state, and for this reason less likely to inspire trust than Bell.

Finally, there is the question of just how much water the

90 J. French to F. B. Crowninshield, 3 Mar. 1859.
91 Martin A. Haynes, *Historical Sketches of Lakeport New Hampshire Formerly Lake Village, Now the Sixth Ward of Laconia* (Lakeport, N.H., 1915), 11.
92 *One Hundred Anniversary of the Incorporation of Gilford*, 24–5.
93 For a brief biography of French, see Courier-Citizen Company, *Illustrated History of Lowell and Vicinity* (Lowell, 1897), 557–9.

company actually drew out of the New Hampshire lakes. Perhaps increasing management of the water during the 1850s aggravated an already tense situation. Typically, the period from late July into autumn was when the Lowell and Lawrence factories often fell short of water. In the years from 1852 to 1865, James Francis calculated that the company drew water from the New Hampshire lakes an average of 52.5 days during the summer and fall months. In 1856, the lakes were used for only 15 days. The following year, no water was drawn. But in 1859, the year of the riot, the company tapped the New Hampshire lakes for the longest period on record between 1852 and 1865, an extraordinary 117 days.[94]

It would be easy to seize on this fact alone in explaining the rebellion at Lake Village. Here was a year when the company created friction by constantly manipulating the water, holding it back when conditions at the Massachusetts mills improved, and sending it forth as drier conditions prevailed. On the surface, this sounds persuasive. Yet a closer look reveals some problems with this interpretation. In 1859, the Lake Company did not begin drawing water from the lakes until 16 August, reasonably late in the season (26 August 1858 was the latest date at which it began drawing water between 1852 and 1865).[95] This was probably late enough to avoid conflict, for example, with meadow owners harvesting hay downstream along the Winnipesaukee River. In all likelihood, the company's water use during the summer approximated what it had used in the past. The year was probably exceptional in that the company continued to draw water well into the autumn. The attack on the Lake Village Dam, however, occurred at the beginning of the fall season, too early to be attributed solely to the company's water management.

Thus far we have examined the general sources of conflict between the company and the community. But to fully explain why a group of people attacked the company's dam, a deeper level of analysis is needed. After the dramatic events of 28 September 1859, the local newspaper reported:

> There has been, and is now quite a strong feeling among not a few people of Lake Village, and indeed the whole county round the

---

[94] J. Francis to Directors, 10 July 1866.
[95] Ibid.; "Dates of commencing to draw water from the Lake Company's Reservoirs in New Hampshire, after the spring freshets, in response to calls from Lowell or Lawrence," 27 Aug. 1877, vol. A-46, file 274, PLC Papers.

Lake, against the course which the Lake Company as it is called, has seen fit to pursue. This feeling, however is not participated in by the whole community by any means. But it was necessary to have some one to go ahead and stir up this feeling – somewhat latent – at least quiescent; and the men at last were found.[96]

Who were these men? It is hard to say who the newspaper had in mind. But Josiah French believed James Worster and George Young had instigated and led the attack.[97] Let us turn to their motives.

When they devised their plan to damage the dam, Worster and Young were in their fifties. They both were also blacksmiths at an earlier time in their lives. Neither of them seems to have experienced much sustained economic success, their fortunes appearing somewhat checkered. Young listed his occupation as laborer in the 1850 census (his name not appearing in the census taken a decade later).[98] When he died in 1870, he owned no real estate.[99] Yet at one point, as we shall see, he managed to rent a factory from the Lake Company for several hundred dollars a year. Worster owned no real estate according to the 1850 census.[100] Although he valued his property holdings in 1860 at \$2,500, much of this land was probably either heavily mortgaged or leased. Still, over the course of his life, he was able to get his hands on some land – more than can be said for others involved in the attack on the dam.

James Worster had experience with breaking down dams. In December 1847, while still living in Dover, New Hampshire, he tore off an abutment, chopped down planking, and removed stone from a dam across the Salmon Falls River in Somersworth, New Hampshire.[101] The dam and factories belonged to the Great Falls Manufacturing Company, a Boston Associates' venture since the 1830s.[102] Claiming damage to land he leased,

96 "Trouble at Lake Village," *Winnipisaukee Gazette*, 1 Oct. 1859.

97 J. French to J. Francis, 21 Sept. 1860, vol. A-38, file 205, PLC Papers.

98 Manuscript Schedules, U.S. Census of Population, Meredith, N.H., 1850.

99 Young died intestate. See case no. 7820, Merrimack County Probate Registry, Concord, N.H.

100 Manuscript Schedules, U.S. Census of Population, Dover, N.H., 1850.

101 See *Great Falls Manufacturing Company* v. *James Worster* (1854), vol. 17:691–6, NHSCN. Worster may have made an earlier attempt to destroy a dam owned by the Great Falls Company. See *Great Falls Co.* v. *Worster*, 15 N.H. 412 (1844). The facts of the case refer to the defendant simply as Worster; no first name is given. Unfortunately, there are no supreme court notes for the case, making conclusive identification difficult.

102 J. D. Van Slyck, *Representatives of New England Manufacturers* (Boston, 1879), 573–5.

Worster sought to abate the nuisance himself – an action that was legal at the time. The Great Falls Company appealed to the New Hampshire Superior Court of Judicature to issue an injunction barring Worster from doing any further damage. In July 1853, the court granted the request.[103]

Worster's history with the Lake Company went back ten years before the 1859 riot. In 1849, his daughter, Adeline E. Worster, took the company to court for flooding her land in Tuftonborough, on the northeast side of Lake Winnipesaukee. She owned the land jointly with her father and claimed the Lake Company's dam at Lake Village had raised the water in the lake and damaged the property. The Lake Company demurred, a move that led to the dismissal of the case in 1852.[104]

Meanwhile, in the period from 1849 to 1853, James Worster made several land transactions: He leased a parcel of meadowland in Sanbornton, a farm bordering Paugus Bay in Gilford, and had a mortgage for a third share of Rattlesnake Island in Lake Winnipesaukee.[105] It is hard to say precisely why he chose these particular tracts of land. Yet one thing is certain: The land seemed destined to bring him into conflict with the Lake Company. On 14 April 1853, Worster threatened to destroy the company's dam at Lake Village, claiming it injured land he owned and leased in neighboring towns.[106]

To protect its property, the Lake Company sought an injunction from the superior court in 1854. At the time, James Bell was still the company's agent and his brother, Samuel D. Bell, was seated on the court. Writing a colleague on the judiciary, Samuel Bell expressed his reluctance to interfere in the case given his brother's stake in the outcome.[107] As to his actual role, we cannot say. But the court did issue an injunction preventing Worster from meddling with the Lake Company's dam at Lake Village in 1855.[108] Violation of an injunction can lead to a contempt charge and possible jail for the offender.

---

103 *Great Falls* v. *Worster* (1854), vol. 17: 770–2, NHSCN.
104 *Worster* v. *Winnipiseogee Lake Co.*, 25 N.H. 525 (1852); *Adeline E. Worster* v. *Winipissiogee Lake Cotton and Woolen Manufacturing Company* (1852), vol. 12:365, NHSCN.
105 J. Bamford to J. Worster, 18 Oct. 1849; G. Young to J. Worster, 8 Oct. 1851; J. Worster with E. and W. Barker, 23 June 1851.
106 This information is found in *Winnipissiogee Lake Co.* v. *Worster*, 29 N.H. 433, 436–7 (1854).
107 S. D. Bell to Eastman, 6 Oct. 1854, *Winnipissiogee Lake Company* v. *James Worster* (1854), vol. 23:833, NHSCN.
108 Worster submitted a motion to have the company's case for an injunction dis-

Barred from tampering with the dams at Lake Village and Somersworth, Worster moved to Concord, New Hampshire. For the moment, he kept out of the way of the Lake Company, but he had in no way given up his fight. Between 1856 and 1858, Worster obtained property in Hooksett, New Hampshire – land bordering the Merrimack River. The land, which was probably prone to flooding, lay upstream from the Amoskeag Company's dam in Manchester.[109] Once again, Worster seemed to be inviting conflict with a Boston Associates' venture.

At half past six on the morning of 7 March 1859, Worster and another person appeared at the Amoskeag Company's dam.[110] The watchman on duty spotted them and ordered them to leave. They refused to go and, after having words, the watchman pitched a piece of ice at them. A fight broke out and Worster was knocked down three times before he left the dam, sending for a doctor to dress his injured nose.

The following month, Ezekiel Straw, the Amoskeag Company's agent, retaliated against Worster. Straw obtained an unpaid note in Worster's name, and managed to have the sheriff attach his property in Hooksett for nonpayment. Soon after, John Harvey, the person from whom Worster had obtained the land in Hooksett, lost a case with the Amoskeag Company for the flooding of his property. In July 1859, Harvey, Worster, Worster's son, and four others made their way to the Amoskeag Company's dam. This time Straw had learned beforehand of their plans. He arrived in time to have the sheriff arrest them for conspiracy as they attempted to tear the flashboards off the dam.

The summer before the attack on the Lake Village Dam, James Worster was on trial for attempting to destroy the Amoskeag Company's dam. In a little more than ten years, he

missed. He argued that the court should not have jurisdiction over the case. The court disagreed and found for the company. A court of equity, the opinion (not written by Samuel Bell) explains, "will not interfere in cases of nuisances, trespasses and the like, where the parties can settle their rights in a court of law, unless it shall appear that irreparable mischief will be done by withholding the process." See *Lake Co. v. Worster,* 29 N.H. at 449. Evidence that documents the issuing of the injunction can be found in the court records for *State v. Worster* (1863), which are included with the papers for *State v. Young* (1860), Trial Term, file 460, N.H. SJC Papers, Belknap County Courthouse, Laconia, N.H.

109 J. Harvey to J. Worster, 26 Sept. 1856, bk. 135:7; J. Harvey to J. Worster, 3 June 1858, bk. 144:490, Merrimack County Registry of Deeds.

110 This paragraph and the one that follows are based on Journal of Ezekiel Straw, 7 Mar., 19, 21 Apr., 7 June, 27, 28 July 1859, D-3, box 6, vol. 2, AMC Papers, MHA.

had made threats or actually damaged parts of three major water control projects in the state of New Hampshire. His behavior expressed sheer determination to interfere with the Boston Associates' plans to manage the region's waters. Justice Samuel Bell once described Worster as "so much a man of one idea, that it is of no use to talk with him."[111] Since Bell did not elaborate, Worster's motives are hard to determine. But it is clear that Worster was indeed a man of enormous resolution. He seemed quick to accuse, tenacious, and willing to do battle with powerful companies. Perhaps his unfaltering determination struck some as charismatic, drawing them toward rebellion. Or maybe they simply had enough reasons of their own.

George Young's confrontations with the Lake Company were somewhat less dramatic. In 1851, Young leased the company's cotton factory on the Winnipesaukee River in Lake Village. For four hundred dollars per year, Young got the factory, a machine shop, blacksmith's shop, counting room, machinery, and waterpower to produce textiles. He did not get the right to control the water in the river and mill pond upstream from the factory. The Lake Company reserved that privilege for itself.[112]

By 1853, Young owed the Lake Company $500 in overdue rent. The company brought suit to recover the money, while the sheriff was called on to attach Young's assets.[113] Seven years later, in 1860, a jury returned a verdict for the company awarding it $373 in damages plus court costs.[114] The case may suggest why Young felt compelled to attack the dam the year before. Young apparently charged the company with not allowing him enough water to properly operate the machinery in his mill. Moses Sargent, who leased the woolen mill from the Lake Company on the same flume below Young, testified against his

---

[111] S. D. Bell to C. Bell, 8 Aug. 1853, *Great Falls* v. *Worster* (1854), vol. 17:779, NHSCN.

[112] *Winipisiogee Lake Cotton and Woolen Manufacturing Company* v. *George W. Young* (1854), vol. 23:820–1, NHSCN. A record of the court of common pleas dated 1854 exists in this volume. Originally, the Lake Co. brought its case in this court, but when the case finally went to trial in 1860, it took place before the SJC. The New Hampshire Court of Common Pleas was abolished in 1859, and the SJC took over its cases.

[113] *Lake Co.* v. *Young* (1854), vol. 23:821–2, NHSCN. Young does not seem to have done well financially in 1852. In April and July, he mortgaged two pieces of property in Meredith and obtained $750 in the process. See G. Young and E. Blaisdell, 22 Apr. 1852, bk. 19:225; G. Young and D. C. Webster, 3 July 1852, bk. 19:407, BCRD.

[114] *Lake Co.* v. *Young* (1860), Trial Term Docket Book (Civil), N.H. SJC Papers.

contention. But a rather surprising witness came forth to support Young. Thomas Ham, James Bell's assistant and briefly in charge before the company hired French, took the stand to speak on Young's behalf. After the trial, French expressed satisfaction with the award, especially "after the evidence given by Mr. Bell's *confidential* assistant and agent as he styled himself *Thomas Ham* Esq. who swore on the stand that Mr. Young could not run his mill at speed with less than ten feet head and fall."[115] It is unclear whether Young's charges were true. Ham of course could have been a disgruntled former employee. However, the jury awarded the company less than what it asked, accounting, almost certainly, for its denial of adequate water.[116] Young then may have been a victim of the company's water control policies and had ample reason to feel angry in 1859.

He had another reason as well. It concerned a piece of land with a complicated history, property bordering the eastern shore of Paugus Bay above the Lake Village Dam. The land once belonged to Moses Rowell, who could hardly have imagined so much trouble would someday erupt over it. In 1841, he died, and his property passed on until his granddaughter leased George Young half of this land for a mere five dollars a year in 1851.[117] When Young went to the Lake Company's dam in the fall of 1859, he intended to remove planks to drain water off of this piece of property.

In the year Young leased the land, the Lake Company rebuilt the dam at Lake Village and flooded the property. His decision to rent the land was perhaps meant to extract money from the company for the flooding it did. The month following Young's lease, he in turn rented the land to Worster.[118] When George

---

[115] J. French to J. Francis, 10 Feb. 1860, vol. A-38, file 205, PLC Papers.

[116] Ibid. Young filed a set-off in the case for $1,100. In a case in which a plaintiff sues a defendant for damages, a defendant can file a set-off, which is a sum of money that the defendant holds is due him from the plaintiff. Since the company was not awarded the full amount of the damages it was seeking, the jury must have believed Young's claim that the company had indeed denied him waterpower.

[117] J. Plumer to G. Young, 27 Sept. 1851, bk. 18:75, BCRD. See *Eastman v. Plumer*, 46 N.H. 464, 476 (1866), for a convenient summary of the deeds and other transactions involving this controversial piece of land. Jacob Rowell, Moses Rowell's son, had two married daughters who, with their husbands, divided the Rowell farm. James H. Plumer and his wife Abby (Jacob Rowell's daughter) took the forty acres bordering Paugus Bay. See J. H. Plumer and A. P. Plumer to S. S. Ayer and M. Ayer, 20 Aug. 1851, bk. 17:585; Ayer to Plumer, 20 Aug. 1851, bk. 17:586, BCRD.

[118] G. Young to J. Worster, 8 Oct. 1851, bk. 18:129, BCRD.

Young and James Worster met is unclear. We know little about their relationship with one another except that in some respects – age, occupation, finances – their lives were comparable. But when Young leased Worster the land, the two came together, bringing their shared hostility to a company each perceived had wronged them. A lot more than land remained bound up in this property. It may have represented a source of income to Young; he claimed to have worked the soil. But it also signified a means of revenge, a way of getting back at the Lake Company.

The company, however, foiled their scheme. By 15 August 1859, French arranged to buy the land it flooded adjoining Paugus Bay from the owners of the property. In another transaction, the company acquired the land rented to Young with title to the lease.[119] George Young now had a new landlord. The Lake Company actually had an easement – a right to flood some of this land – dating from 1845.[120] But the additional flooding caused by the reconstruction of the dam in 1851 (built just a half inch higher but significantly tighter than the old one) forced the company to renegotiate its right to flood. In contrast to the easement, the deeds signed in 1859 conveyed the land itself, giving the company complete reign over the property. When French wrote James Francis to give him the news, a great burden evidently had been lifted. As he put it, "I 'breathe much freer and easier.' "[121] Frustrated by this turn of events, Young and Worster notified French – the same day the company took control of the property – that four feet of water stood on their land.[122]

Of the other rioters, two of them, Thomas Plumer and Augustus Merrill, had dealings with the Lake Company before the attack. Plumer had sold the company in 1845 – before its takeover by the Boston Associates – the right to maintain the Lake Village Dam at a height that flooded his land in Gilford.[123]

---

[119] A. P. Plumer and J. Plumer to Lake Co., 13 Aug. 1859, bk. 32:528; J. Plumer to Lake Co., 13 Aug. 1859, bk. 32:538; J. Plumer to Lake Co., 15 Aug. 1859, bk. 32:542, BCRD. The first deed sold four acres of land bordering Paugus Bay for $1,000. The second one conveyed the right to flow a piece of land known as the Langly Farm. And the third one sold the property Young leased and all title and interest in the lease.

[120] The Lake Company had paid Jacob Rowell ten dollars for the right to flood this property. See J. Rowell to Lake Co., 31 July 1845, bk. 7:337, BCRD.

[121] J. French to J. Francis, 15 Aug. 1859, vol. A-38, file 205, PLC Papers.

[122] Ibid., 25 Aug. 1859.

[123] T. J. Plumer to Lake Co., 31 July 1845, bk. 7:338, BCRD. Plumer sold this right for thirty dollars. In the years after the agreement, the Lake Company embarked on its

Augustus Merrill had a mortgage with the company on a small piece of land in Lake Village.[124] No evidence exists of trouble over the mortgage. But Merrill may have been persuaded to look unfavorably upon the company since he and George Young, according to the census, were once neighbors in Meredith.[125]

The picture we have here is of two people, Worster and Young, out for revenge, urging on a small group of men, most of whom were economically marginal, a few of more substantial means. Frustrated in their dealings with the company, Worster and Young were able to tap into a prevailing current of discontent. In part, we might speculate, that discontent stemmed from the broader economic transformation of the region that did not benefit everyone equally. Whatever bitter feelings there were in this regard were aggravated by the intrusion of out-of-state companies trying to profit from New Hampshire's vast waters. The Lake Company's monopoly over the area's water wealth probably seemed especially galling because it benefited the financial interests of factories in another state. To those of little means, the company's scheme may well have seemed outrageous, reason enough perhaps to drive them to attack the dam.

With their axes and bars, the rioters vented their rage on the most powerful and important symbol of the Lake Company's presence in the region. That dam was more than a mere contrivance for regulating water. It stood as a visible manifestation of the company's elaborate attempt to control water, to appropriate it in the interests of those foreign to this local community. The dam controlled the destiny of the water and represented power over nature. But it also presided, less directly, over the fate of the region and those who lived there. When those axes came down on it that day, it was as if those men were trying to reexert control over the region's most spectacular resource, and ultimately over their own lives.

grand plan for controlling the region's waters. Although no direct evidence exists, it is possible that Plumer felt taken in by the company.

[124] Lake Co. to A. Merrill, 5 Oct. 1858, bk. 31:287; A. Merrill to Lake Co., 7 Oct. 1858, bk. 31:214, BCRD. Merrill purchased land in Lake Village from the Lake Co. and then took out a mortgage for seventy-three dollars on the land.

[125] Manuscript Schedules, U.S. Census of Population, Meredith, N.H., 1850.

5

THE LAW OF WATER

George Young returned home from the 1859 riot at the Lake Company's dam to write a letter to the local newspaper. The letter professed his motives in the affair. Young informed readers of his attempts to settle with Josiah French and the company over the flooding of his land on Paugus Bay. He felt compelled to damage the dam,

> . . . if by doing so the Company would arrest me and thereby bring the matter before a proper tribunal for discussion.
>
> But no – Mr. French, in behalf of the Company, would risk no such experiment. Then I offered them the privilege of sending a man to interfere with me, so that I could arrest him, thereby becoming plaintiff myself. But this was no more an effectual inducement than the other. Again, I offered to submit the matter to two of the justices of the Supreme Court, but this offer also was without avail; the truth being that neither Mr. French nor the Company whom he represents, are willing to have a judicial decision upon the question of rights of individuals, for the Company in this matter have no rights but might.

Having failed to convince the company to involve a judicial authority, Young claimed he had only one choice left him. "I have at length peaceably and quietly attempted," Young wrote, "to use the only remaining remedy which the law confers upon me, viz: to peaceably and quietly abate the nuisance with my own hands."[1]

Young would have his day in court. Four legal cases resulted from the attack on the Lake Village Dam. Of these, three were for criminal charges, all stemming from the actual fighting – cases that dealt with the actions and events of the past. The fourth case sought to shape the future. It was brought by the Lake Company to enjoin the rioters from ever again interfering with the company's dam. That case had immense significance

[1] Letter to the Editor, *Winnipisaukee Gazette*, 8 Oct. 1859.

for the company, and it will be the main focus of this chapter. Overall, the company won three of the four cases, including its effort to forbid any further interference with its dam. One case appears to have ended in a hung jury.

As Young's letter suggests, the law structured the actions of both the Lake Company and those who opposed it. The conflict at the dam took place within an established legal framework. Young believed, as he argued in the newspaper, that he acted within the law when he went peaceably – although it clearly turned out to be anything but – to the dam to abate the nuisance. Did the law actually afford him the right to remove the dam because it flooded the land he leased? Only by turning to the substance of nineteenth-century New Hampshire water law can the question be answered. It remains to explore how the law regulated the use of water, how it adjudicated the conflicts that arose, and who benefited as a result.

Revenge seldom fails to bring about reprisal, and it is no surprise to see the Lake Company pursue James Worster. When Worster went to wreck the company's dam on 28 September 1859, he really took his chances. Four years earlier, the company had secured an injunction to prevent him from interfering with its property. Now his presence during the riot provoked the company to take further action. Two weeks after the riot, Josiah French swore out a complaint against Worster. Worster, he charged, had violated the injunction against him by attempting once again to demolish the Lake Company's dam and should be prosecuted for contempt.[2] Thus began a protracted legal proceeding that haunted James Worster for several years.

Worster denied the charges.[3] Beginning in the spring of 1860, depositions were taken to determine whether or not he had violated the injunction.[4] Nathaniel Folsom, Jr., had earlier vouched that he had helped George Young remove part of the dam and that Worster had played no part in the affair.[5] There must have been a great number of depositions taken, for the

[2] Petition of the Lake Co., 14 Oct. 1859, *State v. Worster* (1863), Trial Term, file 460, N.H. SJC Papers, Belknap County Courthouse, Laconia, N.H.

[3] Statement of J. Worster, 9 Dec. 1859, *State v. Worster* (1863), Trial Term, file 460, N.H. SJC Papers.

[4] J. French to J. Francis, 11 May 1860, A-38, file 205, PLC Papers, BL.

[5] Deposition of N. Folsom, Jr., 10 Dec. 1859, *State v. Worster* (1863), Trial Term, file 460, N.H. SJC Papers.

process dragged on into the summer.[6] The weather got hotter and so did Worster's temper. In late July, Worster tried to assault the Lake Company's attorney, Thomas J. Whipple, as he took his statement. Instead, according to French, the glass door to Whipple's office was smashed and "Worster got his face blacked by receiving the contents of a very heavy inkstand, and in some way unknown came to the floor with such a crash as to cause the occupants of the store below to rush up stairs [*sic*] to see what had happened."[7]

Before his role in the attack had been settled, Worster was jailed on another offense. Sometime before the autumn of 1859, Worster physically resisted the Merrimack County sheriff in a dispute over a stolen horse. In 1860, Worster pleaded guilty to the charge and was sentenced to thirty days in jail the following year.[8] When he heard the news, French wrote: "He ought to be in jail or in an Insane Asylum. Perhaps the latter place would be the most humane and proper."[9] Worster was eventually released but returned to jail two years later after being convicted of contempt for his part in the 1859 attack on the dam. The sentence included three months of incarceration and a five-hundred-dollar fine.[10] The Lake Company had finally succeeded. James Worster was to cause the company no further trouble.

Two additional criminal cases resulted from the attack on the dam at Lake Village. As the company pursued Worster, it also took action against the other major participants in the riot. French brought his case before Stephen C. Lyford, the local justice of the peace, who was responsible for hearing charges of public disorder. It is an odd coincidence that Lyford (and Nathan Batchelder) had in 1829 built the dam at Lake Village[11]

6 J. French to J. Francis, 13 July 1860, A-38, file 205, PLC Papers.
7 J. French to F. B. Crowninshield, 25 July 1860, A-38, file 205, PLC Papers. Worster was badly injured, for French wrote: "Blood flowed from a cut in Worster's head which made such a variety of colors as to make it look rather ludicrous. His right shoulder was so injured, as he said that he could not write his interrogatories. . . . I felt sorry to hear such a thing take place and that Col. Whipple should have his office disgraced and his door smashed to pieces."
8 SJC Reports, *New Hampshire Patriot*, 29 Aug. and 5 Sept. 1860; J. French to J. Francis, 24 Sept. 1861, A-38, file 211, PLC Papers.
9 J. French to J. Francis, 24 Sept. 1861.
10 Verdict, *State v. Worster* (1863), Trial Term, file 460, N.H. SJC Papers. After serving the three months, Worster was still unable to pay the entire fine and pleaded with the SJC to dismiss $155. No record exists of the court's response. See J. Worster to N.H. SJC, 3 Dec. 1863, *State v. Worster*, ibid.
11 See the facts in *Winnipiseogee Lake Co. v. Young*, 40 N.H. 420–1 (1860).

– the same dam the Lake Company bought and reconstructed in 1851. One wonders what passed through Lyford's mind as he determined the fate of those who threatened to demolish his construction project. French charged Young, Worster, Augustus Merrill, Nathaniel Folsom, Jr., and seven others with rioting at the dam and attempting to kill him by pushing him off it into the water below.[12] In October 1859, Lyford conducted a brief hearing and held each of the defendants on two-hundred-dollar bail pending their appearance at the next trial term of the supreme judicial court.[13]

The case went to trial in 1860. On 4 October, after the arguments of both sides were presumably heard, the presiding judge sent the jury out to deliberate. French felt pessimistic about the company's chances in the case. He hoped of course to see the rioters convicted, but "the uncertainty of getting twelve men to agree upon a verdict of guilty where there are so many interested and hoping to devise some advantage from such attempts, is very great."[14] French's doubts proved well founded. The trial ended in a hung jury. Although there is some evidence of a retrial, none of the evidence – including the actual court files – contains a verdict. And in all likelihood, one was never delivered.[15]

There was, however, a verdict in Young's countersuit against French and his assistant, Augustus Owen – the third and final criminal case resulting from the attack. Young sued them for assaulting him with an iron bar during the riot. French and Owen claimed their assault constituted an act of self-defense. In their view, they were merely protecting the dam from Young, who had unlawfully entered the company's property to cause damage. In 1861, the jury returned a judgment for the defendants.[16] George Young's day in court had not gone quite the way he had hoped. But he would get another chance.

The Lake Company had a much more significant objective than to prosecute rioters for past offenses. To continue benefit-

[12] S. C. Lyford's Notes on the Appearance of J. French, n.d., *State v. Young* (1860), Trial Term, file 460, N.H. SJC Papers.
[13] *Laconia Democrat,* 21 Oct. 1859.
[14] J. French to J. Francis, 21 Sept. 1860, A-38, file 205, PLC Papers.
[15] SJC Report, *Winnipisaukee Gazette,* 13 Oct. 1860. Also see J. French, Annual Report of Lake Co., 17 May 1861, item 204, EC Papers, MATH. The verdict, according to French, was ten for conviction and two for acquittal.
[16] Pleadings and Verdict, *Young v. French* (1860), Trial Term, file 4938, N.H. SJC Papers.

ing from the waters of Lake Winnipesaukee, it needed security from any further harm to the dam at Lake Village. Almost immediately after the 1859 incident, French petitioned the supreme judicial court (SJC) for an injunction naming eight participants in the riot, including Worster, Young, Nathaniel Folsom, Jr., Thomas Plumer, and Peaslee Folsom (four individuals indicted on criminal charges for riot were not named). The SJC, which alone had the power to grant injunctions, issued a temporary restraining order on 5 October 1859.[17] The defendants tried to block the order by demurring. Ultimately, the case for a permanent injunction was heard at the June 1860 session of the SJC.

In 1860, New Hampshire's court system had two levels, a set of lower police courts and justices of the peace, and the higher SJC. Cases for breach of the peace and other minor criminal proceedings generally were handled by local justices of the peace.[18] As late as 1859, New Hampshire also had a court of common pleas, before which a variety of civil cases were heard. In that year, the legislature abolished this court and reassigned its duties to the SJC. That court met in two different capacities, as a trial court and as a court of appeal. At its trial term, juries were empowered to determine facts and render decisions (as was true in the three cases discussed above). But when the SJC met in its capacity as a court of appeal, no juries were present. And for the most part, the court did not decide the facts of a case. Their task was in some ways a more lofty one: They heard arguments that allowed them to interpret how the law should be applied. On occasion, the court during this term also exercised its authority as a court of equity. Ordinarily, the SJC administered justice by adhering strictly to the principles laid down by the common law, awarding damages, if need be, based on what this body of law allowed. Cases in equity, however, involved a different set of remedies. During its equity term, the court had the power to issue injunctions to redress a situation.[19]

17 *Writ of Injunction v. G. W. Young & Others*, 5 Oct. 1859, item 204, EC Papers. Justice Asa Fowler signed the order. Also see *Winnipisaukee Gazette*, 8 Oct. 1859.
18 The Lake Company first turned to its justice of the peace after the attack on the dam, but when the company brought charges of riot – a state offense – the state became involved and the case moved to the higher court.
19 For a brief discussion of the New Hampshire court system in the early nineteenth century, see Alfred S. Konefsky and Andrew J. King, eds., *Legal Papers, The New Hampshire Practice*, The Papers of Daniel Webster, vol. 1 (Hanover, N.H., 1982), 63–5. The court was remodeled in 1813, 1855, and 1859. In particular, see Act of 14

In the years from 1850 to 1870, companies controlled by the Boston Associates came before this court concerning water-related cases no less than thirteen times. The Great Falls Manufacturing Company and the Amoskeag Company each appeared four times, while the Lake Company showed up on five occasions.[20] The bulk of the cases involved the flooding of land, including the Lake Company's 1860 suit to prevent any further destruction of its dam. This particular case was of enormous importance to the Lake Company. A victory would give it the legal leverage it needed to continue controlling Lake Winnipesaukee's water to benefit the factories in Lowell and Lawrence. The outcome of the case was significant and the reasons for the specific decision even more so. Since it dealt with water and the flooding of land, it cannot be properly understood without first exploring the history of New Hampshire water law.

"Running water is not in its nature private property." Those are the words of an English jurist writing in the early nineteenth century.[21] Although most American jurists recognized the validity of such a statement, they spent much time and effort trying to finesse this apparent truism. During the antebellum years, American water law cases sought to reconcile the fluid, elusive nature of water with a social system that had private property at its heart. The legal doctrines that evolved are fraught with confusion and ambiguity, but the general direction of the law by the 1860s is clear. Over the course of the nineteenth century, the law moved to define water's place in the process of industrial production. A new conception of water as a form of property emerged – a new understanding of how this resource could be used to suit the needs of a growing economy.

July 1855, ch. 1659, [1855–61] N.H. Laws 1538; Act of 28 June 1859, ch. 2211, [1855–61] N.H. Laws 2088.

20 *Great Falls Mfg. Co. v. Worster*, 23 N.H. 462 (1851); *Worster v. Winnipiseogee Lake Co.*, 25 N.H. 525 (1852); *Winnipissiogee Lake Co. v. Worster*, 29 N.H. 433 (1854); *Coe v. Winnepisiogee Lake Cotton & Woolen Mfg. Co.*, 37 N.H. 254 (1858); *Pray v. Great Falls Mfg. Co.*, 38 N.H. 442 (1859); *Winnipiseogee Lake Co. v. Young*, 40 N.H. 420 (1860); *Worster v. Great Falls Mfg. Co.*, 41 N.H. 16 (1860); *Hooksett v. Amoskeag Mfg. Co.*, 44 N.H. 105 (1862); *Eastman v. Amoskeag Mfg. Co.*, 44 N.H. 143 (1862); *Amoskeag Mfg. Co. v. Goodale*, 46 N.H. 53 (1865); *Winnipisseogee Lake Cotton & Woolen Mfg. Co. v. Perley*, 46 N.H. 83 (1865); *Eastman v. Amoskeag Mfg. Co.*, 47 N.H. 71 (1866); *Great Falls Mfg. Co. v. Fernald*, 47 N.H. 444 (1867).

21 *Williams v. Morland*, 2 Barn. & Cres. 910, 107 Eng. Rep. 620 (K.B. 1824). The statement was made by Judge Holroyd.

The evolution of the mill acts in Massachusetts, as noted in Chapter 1, marked a shift by the late eighteenth century toward a more instrumental vision of water. In New Hampshire, the broad contours of nineteenth-century water law, with some exceptions, suggest a similar drive to value water primarily as a source of economic potential.

Before the nineteenth century, the common law had taken a much different attitude toward water. *Aqua currit et debet currere, ut currere solebat*: Literally, water flows and ought to flow, as it has customarily flowed. It was a simple principle that formed the foundation for the common law of watercourses.[22] That law was premised on the inherent good of a natural flow, of allowing water to go forth as nature had ordained. The general rule, as articulated by the English jurist Lord Ellenborough in 1805, was that "independent of any particular enjoyment used to be had by another, every man has a right to have the advantage of a flow of water in his own land without diminution or alteration." Lord Ellenborough's words ring forth in the American water law cases – indeed his opinion was often cited by jurists seeking to justify a natural-flow rule – into the nineteenth century.[23] But his statement carries with it an implicit assumption. Clearly there could be no strict adherence to the natural-flow rule without almost completely denying people the use of this valuable resource. The law had to allow at least some minimal appropriation of water.

The natural-flow rule sanctioned a restricted range of justifiable water use. To avail oneself of water for domestic purposes or for husbandry seems never to have been challenged in either Britain or America. An upstream riparian proprietor could take water to drink, wash, and water cattle even though such uses might prevent owners further downstream from doing the same.[24] Jurists in the eighteenth century, however, were wary of the use of water to irrigate or propel machinery. Such uses required significantly more water and were less crucial than using the water for domestic needs. (When the Scottish jurist Lord Kames enumerated the preferences for water use, run-

---

[22] Joseph K. Angell, *A Treatise on the Law of Watercourses*, 5th ed. (Boston, 1854), 94.

[23] *Bealey v. Shaw*, 6 East 208, 102 Eng. Rep. 1266 (K.B. 1805). It was common for American jurists to pay homage to the natural-flow rule by citing this case. See, e.g., *Arnold v. Foot*, 12 Wend. 330, 332 (N.Y. 1834).

[24] Francis H. Bohlen, "The Rule in *Rylands v. Fletcher*," *University of Pennsylvania Law Review* 59 (1911): 321–2, n. 25.

ning machinery with it ranked last.)[25] Thus the "natural" use of water meant tapping it to sustain life, both human and animal.[26] Beyond that, its consumption for other purposes was subject to a stricter legal qualification – *sic utere tuo ut alienum non laedas*, that is, use your own [property] so as not to injure that of another – invoked to limit the competing uses of water to protect the property of all.[27] As propounded in eighteenth-century America, the common law placed strict limits on the appropriation of water. It was premised on a narrow conception of water's productive potential.

In the fifty years between 1820 and 1870, New Hampshire's highest court tried to make sense of this common-law legacy. A range of water cases were tried before it, but there were three main types. Most of the cases involved an action by an upstream landowner for flooding to land by a dam erected downstream. There were also cases in which a downstream owner sued an upstream proprietor for diversion. A handful of cases concerned an upstream mill owner who took his downstream counterpart to court for flowing water back and preventing his waterwheels from turning. The New Hampshire court strove to reconcile the aspects of the common law opposed to the most productive use of water with the need for economic development. In general, the cases conformed to the broader trend in nineteenth-century property law described by Morton Horwitz. The emerging requirements of the developing American economy impelled a change in the concept of property, "from a static agrarian conception entitling an owner to undisturbed enjoyment, to a dynamic, instrumental, and more abstract view of property that emphasized the newly paramount virtues of productive use and development."[28] In the twenty years after midcentury, this more instrumental vision of property emerged in New Hampshire water law. Although this was clearly the general direction of the law, the specifics of New Hampshire's water law differed from other states. This will be most apparent when the state's mill acts are examined. But first let us consider

---

[25] See the argument made in *Merritt v. Brinkerhoff*, 17 Johns. 306, 316 (N.Y. 1820), in which Kames's views are summarized. Also see Morton J. Horwitz, *The Transformation of American Law, 1780–1860* (Cambridge, Mass., 1977), 32, 36.

[26] For a strict delineation of the natural-flow rule by an American court, see *Evans v. Merriweather*, 4 Ill. 492 (1842).

[27] T. E. Lauer, "Reflections on Riparianism," *Missouri Law Review* 35 (1970): 1, n. 2; Horwitz, *Transformation of American Law*, 32.

[28] Horwitz, *Transformation of American Law*, 31.

the legal rules applied by the courts to the manipulation of water.

Before the middle of the nineteenth century, New Hampshire's highest court articulated a doctrine of water regulation that tended to minimize legal injury to property. The opinions encouraged riparian proprietors to use water in ways that would not infringe on the legal rights of others to do the same. Consensus existed, for example, on a mill owner's right to change the mode or object of water use. But the opinions insisted that such a change not increase the quantity appropriated beyond what an owner was legally entitled. A mill owner could introduce new machinery into an establishment so long as the usual amount of water continued to run downstream. Changes to the machinery in a mill, an opinion written in 1820 expounds, "are in general *damnum absque injuria* [damage without legal injury]: and it is doubtless fortunate for the public that there is no check upon them, unless they actually divert a part of the water from those below."[29] Similarly, the law protected landed property from flooding by milldams and other obstructions by presuming such flowage to be an act of damage.[30] The law generally regulated water in a way that failed to maximize its productive potential. The substance of New Hampshire water law by the early nineteenth century was still less instrumental, less devoted to the dictates of production than what emerged later.[31]

Yet there were some clear indications by 1830 of a new, more dynamic conception of property. Before the nineteenth century, it was possible to attain a legal interest in water merely by using it for a long period of time. A riparian owner could exclude others from using water simply by establishing that the water had been appropriated some time ago. This was known as the doctrine of prescription. To establish a prescriptive claim under the old common law, enjoyment of the right must have

---

29 *Bullen v. Runnels*, 2 N.H. 255, 262–3 (1820). Also see *Johnson v. Rand*, 6 N.H. 22–3 (1832) and *Whittier v. Cocheco Mfg. Co.*, 9 N.H. 454, 458 (1838).
30 *Woodman v. Tufts*, 9 N.H. 88, 91 (1837).
31 Horwitz tends to conceptualize certain antidevelopmental aspects of the common law as "preproductive." But surely eighteenth-century New Englanders, for instance, had a productive vision of their relationship with the world. It was a vision, however, that was less instrumental and production oriented than what emerged in the nineteenth century. See Horwitz, *Transformation of American Law*, 39; also see Mark Tushnet, "A Marxist Analysis of American Law," *Marxist Perspectives* 1 (1978): 107.

existed from time immemorial.[32] The doctrine had particular appeal to a tradition-bound culture, for which the past loomed enormously over the future. Prescription justified the status quo in regard to property rights merely because things had always been that way. But as a legal concept, it had outlived its value to American society by the nineteenth century – a point clearly demonstrated in New Hampshire.[33]

In 1820, the New Hampshire court ruled in *Bullen v. Runnels* that occupation of a water privilege for a period of twenty years conferred presumptive evidence of a grant.[34] American courts during the early nineteenth century commonly fixed the length of time necessary for establishing a prescriptive right.[35] After all, the English standard for prescription equating "time immemorial" with a point in the twelfth century seemed irrelevant to a nation settled much later. Still, the New Hampshire court's decision maintained a lenient standard: Mere occupation for twenty years established a grant. How one chose to maintain the privilege, the court seemed to say, was of no consequence. Occupancy alone could be used to demonstrate a legal claim. Little did it matter whether or not the privilege was openly asserted against those who could conceivably challenge such a right.[36]

This standard for evaluating prescriptive claims to water came under attack but ten years later. In *Gilman v. Tilton*, a case in which the defendants' dam caused water to flow back and obstruct the plaintiff's waterwheel, Chief Justice Richardson began by invoking the natural-flow rule.[37] "In general," he wrote, "every man has a right to the use of the water flowing in a stream through his land; and if any one divert the water from its natural channel, or throw it back, so as to deprive him of the use of it, the law will give him redress." He then acknowledged that a riparian proprietor could acquire the right to flood the land of another through "long usage." Yet it seemed to him "well settled, that a man acquires no such right by merely being the first to make use of the water."[38] Simple appropriation of

[32] T. E. Lauer, "The Common Law Background of the Riparian Doctrine," *Missouri Law Review* 28 (1963): 60, 84.

[33] Horwitz, *Transformation of American Law*, 43–7.

[34] *Bullen v. Runnels*, 2 N.H. at 257.

[35] Horwitz, *Transformation of American Law*, 280–1, n. 52.

[36] *Bullen v. Runnels*, 2 N.H. at 255.

[37] *Gilman v. Tilton*, 5 N.H. 231 (1830).

[38] Ibid., 232–3.

the water conferred no right to exclude others from its use. Instead, Richardson proposed "adverse use" as the relevant criterion for presuming a grant. What constituted adverse use? To Richardson it meant that the person claiming the grant had to interfere with another's right, to make clear to others the extent of the claim.[39] By making adverse use, not simply long usage, the standard, Richardson departed from the common law tradition. Under the traditional notion of prescription, those who failed to develop water, letting it languish, could still maintain their legal claim to the resource. Adverse use required the first user of water to develop it or be exposed to the loss of the property right.

Over the next three decades, New Hampshire's highest court followed the concept of adverse use, until by midcentury a consensus had emerged on the doctrine. In *Wallace* v. *Fletcher* (1855), Justice Samuel Bell (brother of the Lake Company's agent, James Bell) wrote an extensive critique of prescription. After reviewing the precedent in English common law for the doctrine, Bell concluded:

> Many cases, in this country, have followed in the tracks of the English decisions, though it is apparent that, in a newly settled country like ours, where to a great extent every thing is of recent date, and the history of our towns, of our roads, farms, mills and dwellings are known, a rule like that adopted in England is in no respect adapted to our situation. On other subjects, the common law has been every where modified, to adapt it to the wants of our community.

Concern for the public good of New Hampshire, in Bell's view, warranted the adoption of "'adverse, exclusive and uninterrupted enjoyment for twenty years'" as the appropriate standard.[40] Bell gave intellectual coherence to the rule of adverse use – a doctrine evolving in the New Hampshire court for twenty-five years.[41] Removing the restraining hand of prescription cleared the way for a more flexible conception of water as a form of property. Moreover, the rule of adverse use had the necessary resilience to suit the needs of the state's developing economy. Adverse use compelled riparian owners to develop their water resources in order to maintain their rights to the water. The doctrine thus encouraged the productive, instrumental use of water.

[39] Ibid., 233–4.
[40] *Wallace* v. *Fletcher*, 30 N.H. 434, 447–8 (1855).
[41] See, e.g., *Odiorne* v. *Lyford*, 9 N.H. 502, 513–14 (1838), and *Watkins* v. *Peck*, 13 N.H. 360, 370 (1843).

The strictures of the common law opposed to economic growth were being challenged in New Hampshire at midcentury. By that time, the doctrine of natural flow had lost a good deal of its validity and was slowly being replaced. Justice Woods in *Gerrish* v. *New Market Man. Co.* (1854), a case in which the plaintiff claimed a mill's reservoir flooded his land, began his decision with a classic statement of the natural-flow rule (i.e., every proprietor along the stream was entitled to use the water without diminution or obstruction). He then qualified this statement. A riparian proprietor could interfere with the natural flow, Woods allowed, for "domestic, agricultural, and manufacturing purposes, provided he uses it in a reasonable manner, and so as to work no actual, material injury or annoyance to others."[42] Such a statement was a long way from the eighteenth-century water cases that recognized limited use of the flow for domestic needs and husbandry. For the court to acknowledge the propriety of using water for manufacturing in the mid-1800s is not surprising. But by grouping industrial and agricultural uses of water – applications demanding significant quantities of water – with domestic ones, the court implicitly sanctioned a certain amount of noncompensable harm to property. The question was, What constituted a reasonable use of water?

Between 1855 and 1863, the phrase "reasonable use" surfaced in at least two water law cases.[43] The idea of reasonable use marked the realization that economic development inevitably caused some amount of damage to the property of others. Although the New Hampshire court had been moving gradually in the direction of this doctrine, the clearest statement of the reasonable-use rule did not appear until 1863, in *Hayes* v. *Waldron*.[44] This case involved the discharge of dust and shavings from a sawmill into a river; the plaintiff claimed damage from the waste deposited on his meadows downstream from the mill. In Justice Bellows's opinion, "some diminution, retardation, or acceleration of the natural current" is consistent with a riparian proprietor's right to use water for domestic, agricultural, and manufacturing purposes provided there was reasonable use. The interpretation of a "reasonable use" was a relative question, he continued, that depended on the size of

[42] *Gerrish* v. *New Market Man. Co.*, 30 N.H. 478, 483 (1854).
[43] See *Tillotson* v. *Smith*, 32 N.H. 90, 95 (1855), and *Bassett* v. *Salisbury Mfg. Co.*, 43 N.H. 569, 577 (1862).
[44] *Hayes* v. *Waldron*, 44 N.H. 580 (1863).

the stream and the way the water was applied – questions making it impossible to give an exact definition. Yet of one thing Bellows felt certain: "The rule is flexible, and suited to the growing and changing wants of communities." Bellows thought it proper to weigh the benefits of different water uses to establish what was reasonable. Determining those parameters required justices to consider "the extent of the benefit to the mill owner, and of inconvenience or injury to others."[45]

This opinion had far-reaching implications. The New Hampshire court had long recognized that a system of reciprocal rights must govern the state's waters. But Bellows's decision went beyond this. His opinion advocated the evaluation of relative efficiencies in deciding water conflicts and was predicated on the notion of maximizing production. Since midcentury, the New Hampshire court had been groping toward a definition of reasonable water use. Bellows provided a persuasive answer: So long as an appropriation of water promoted the welfare of the whole community, it was permissible to take whatever quantity was needed and to cause damage to some individual interests. The language here opened the way for factory owners (and others depending on large amounts of water) to consume significant quantities of the state's water resources if they showed that it benefited the broader community.

The rule of reasonable use, as Bellows noted, was a flexible one. Reasonable water use, he believed, was what judges interpreted to be in the best interests of "communities." But no one of course could say exactly what an entire community thought about a particular water use. This fact left the rule to serve the needs of those who demonstrated that their use of the water benefited everyone, whether it actually did or not. One wonders whether the doctrine of reasonable use was abused by powerful individuals and companies seeking expanded control over the state's water at the expense of other interests. The words of a Vermont jurist, written in response to a similar case, imply that in that state at least, it could be: "Within reasonable limits, those who have a common interest in the use of air and running water, must submit to small inconveniences to afford a *disproportionate* advantage to others [emphasis added]."[46]

Yet the law did not simply allow powerful interests to race

[45] Ibid., 584–5.
[46] *Snow v. Parsons*, 28 Vt. 459, 462 (1856). Also see Horwitz, *Transformation of American Law*, 40–2.

through New Hampshire amassing huge amounts of water with impunity. The doctrine of reasonable use had its qualifications. In the two decades following 1850, as the New Hampshire court struggled to adopt a water policy more consistent with economic growth, it continued to safeguard landed property from flooding by dams. By continuing to protect property from flooding, this policy mitigated, to some extent, the economic potential of the reasonable-use doctrine. A number of cases decided after 1850 strongly assert the right of an upstream riparian landowner to bring action for flowage.[47] In *Bassett* v. *Salisbury Manufacturing Company* (1862), the court invoked the language of reasonable use but held that the rule did not remove the liability for flooding land.[48] Reasonable use was never meant to completely exonerate milldam owners from sending water onto nearby land. The point is made explicit in *Gerrish* v. *Clough* (1868): "The doctrine of reasonable use does not apply to a dam raising the height of the water above the dam [causing the land upstream to flood]."[49] The protection given property flooded by milldams is further demonstrated by examining the history of New Hampshire's mill acts.

George Young claimed he acted within the law when he went to the Lake Company's dam to abate the flooding of his land. For the most part, he was correct. As late as 1859, and for close to a decade beyond, New Hampshire law permitted landowners with property flooded by milldams to take the law into their own hands and remove the nuisance. In nineteenth-century New Hampshire, dam breaking to alleviate the injuries to property was a more popular remedy than in Massachusetts, where a mill act was in force throughout the century. Without recourse to a mill act, New Hampshire factory owners had to be cautious when flooding land lest someone decide to fix the problem by breaking down the dam.

---

47 Writing in 1852, Justice Woods explained, "It has never yet been held, we think, that a riparian proprietor, as such, has the right, by any means, to cause the water in the stream to be thrown back, and to overflow the lands of the proprietor above, upon the stream, to his damage." *Cowles* v. *Kidder,* 24 N.H. 364, 382 (1852). In addition, the court upheld the common-law principle that an action by a landholder for trespass to property can be maintained without actual proof of specific injury. In short, the overflowing of land still was presumed an act of damage in the period from 1850 to 1870. See *Snow* v. *Cowles,* 22 N.H. 296, 302–3 (1851); *Cowles* v. *Kidder,* 24 N.H. at 379; *Tillotson* v. *Smith,* 32 N.H. at 96.

48 *Bassett* v. *Salisbury,* 43 N.H. at 569, 578.

49 *Gerrish* v. *Clough,* 48 N.H. 9, 12 (1868).

A number of remedies existed under the common law for landholders when milldams flooded their land. One measure involved abating the nuisance (i.e., the dam) without appealing to a higher authority.[50] Some New England states – and Massachusetts is a prime example – shed this aspect of the common-law legacy by the early nineteenth century and proved far more congenial to manufacturing interests. The Massachusetts mill acts – which were repassed and amended throughout the first third of the nineteenth century – established a formal procedure for compensating landowners when flooding caused damage. In the process, they removed certain traditional routes, provided by the common law, for seeking relief. Thus a farmer with property flooded by an obstruction could no longer remove the source of the nuisance on his own. The passage of the mill acts precluded such a course of action, insuring the physical safety of dams.[51]

New Hampshire legislated its first mill act in 1718.[52] But the act was probably not interpreted to eliminate the common-law recourse for nuisances to land, and in any event, it was repealed in 1792. Although attempts were made to get such a bill through the legislature (particularly in the 1850s), not until 1868 did the state pass a bill allowing mill owners more freedom to flood land.[53] The explanation lies buried in the political workings of the state. The Democratic party, with its antimanufacturing tendencies, flourished in New Hampshire between the early 1830s and 1855. Hostility to manufacturing interests peaked in the early 1840s: No manufacturing companies were chartered by the state between 1840 and 1842. In part, the severe downturn in the American economy was responsible. Still, New Hampshire, with its legislature controlled by radical Democrats, was hardly a popular place to start a manufacturing company, especially after an act passed in 1842 made stockholders personally liable for the debt of the corporation.[54]

The antidevelopmental bias of the New Hampshire legisla-

---

50 For other avenues of relief, see Horwitz, *Transformation of American Law,* 48.

51 For an extended discussion of the Massachusetts mill acts, see ibid., 47–53; also see William E. Nelson, *Americanization of the Common Law: The Impact of Legal Change on Massachusetts Society, 1760–1830* (Cambridge, Mass., 1975), 159.

52 Act of 14 May 1718, ch. 14, [1702–45] N.H. Laws 265. The act was modeled after the 1713 Massachusetts mill act.

53 Act of 3 July 1868, ch. 20, [1868] N.H. Laws 152.

54 Donald B. Cole, *Jacksonian Democracy in New Hampshire, 1800–1851* (Cambridge, Mass., 1970), 14, 202, 206.

ture became a topic of major concern during the mid-1840s. By this time, news spread of the Boston Associates' intention to build additional textile cities in New England. One report published in the *Manchester American* – a strong supporter of free enterprise – accused the New Hampshire legislature of driving capital out of the state. According to the newspaper, unfavorable legislation – especially the state's unlimited liability law – had twice forced the Boston Associates to give up plans for developing additional industry in the city of Manchester. Instead, the money financed the new textile cities of Saco, Maine, and Lawrence, Massachusetts. "No man can reasonably admit a doubt to cross his mind," the paper wrote in 1845, ". . . that the money which is to build up this 'second Lowell' [Lawrence] would have been invested at 'Manchester, Hooksett, or Bow,' had it not been for radical legislation."[55] Clearly, the state did not offer a promising political climate for the passage of a mill act.

Without a mill act, injured landowners had two choices. They could bring an action for damages and try to recover compensation for their flooded land, or they could take the law upon themselves and abate the nuisance. The latter choice proved especially troubling to manufacturers. In 1844, the New Hampshire SJC heard a case for trespass brought by the Great Falls Manufacturing Company. Attorneys for the plaintiffs tried to convince the court of their own power to "enjoin parties and abate nuisances," making it no longer legal for citizens to remove nuisances on their own. "To allow such a course," they reasoned, "endangers public peace, and puts private rights in extreme jeopardy." The court did not agree. But it did hold that individuals seeking to abate nuisances on their own must "proceed in a reasonable manner." To be reasonable, the court ruled, the affected landholder must cause as little damage as possible when removing the nuisance.[56]

In the spring of 1859, after the earlier attacks on the Amoskeag Company's dam, the legislature considered passing a mill act.[57] The bill found support from the Amoskeag Company's agent, Ezekiel Straw, and may have been drafted by law-

---

55 "Falsehoods refuted," *Manchester American*, 28 Feb. 1845. Similar outcries over the antibusiness sentiment of the New Hampshire legislature can also be found in *New Hampshire Statesman*, 14 Feb. 1845; *New Hampshire Sentinel*, 19 Mar. 1845. For a Democratic perspective on the issue, see "Driving Capital to Maine," *New Hampshire Patriot and State Gazette*, 29 Feb. 1844.

56 *Great Falls v. Worster*, 15 N.H. at 412, 427–8, 438–9.

57 *Journal of the Senate of the State of New Hampshire* (Concord, N.H., 1859), 21 June 1859, 128, 145–6.

yers he hired.[58] In June, the bill reached a third reading in the state senate.[59] New Hampshire had come closer to passing a mill act than at any other time in the nineteenth century. With victory for the bill close at hand, the opposition intervened. John Coe, James Worster, and others appeared at the state-house in Concord and, according to Straw, "by making a great fuss got the nuisance bill [the mill act] recommitted to the Comm. [Senate Judiciary Committee] for a further hearing."[60] Consideration of the bill was postponed to the next session of the legislature where it appears to have died.[61] The bill was blocked by a combination of forces, not the least of which included hostility shown the imperial ambitions of the state's waterpower companies.[62]

At the time the Lake Company sought legal protection for its dam, New Hampshire water law was fraught with tension and ambiguity. Between 1830 and 1870, the state's judiciary was inclined toward the economic development of water, tending to favor its use for manufacturing. There were two important indications of this shift. First, the SJC reinterpreted the doctrine of prescription in a way that emphasized the concept of adverse use. Whereas prescription favored every property owner, even those who might never develop the water, adverse use required owners of riparian lands to control water in order to protect their legal title. An adverse-use rule embodied a more instrumental conception of water and was therefore more consistent with the needs of economic development. Similarly, the court affected the productive potential of water by shifting from a natural-flow rule to the doctrine of reasonable use. The legislature, however, seemed reluctant to sanction instrumental water use when it infringed on the rights of landed property owners. The absence of a mill act for most of the nineteenth century is the strongest evidence of this trend.

* * *

[58] Journal of Ezekiel Straw, 4, 21 June 1859, D-3, box 6, vol. 2, AMC Papers, MHA.
[59] *Journal of the Senate of the State of New Hampshire*, 21 June 1859, 145–6.
[60] Journal of Ezekiel Straw, 22 June 1859.
[61] *Journal of the Senate of the State of New Hampshire*, 22, 24 June 1859, 161, 205–6.
[62] Through amendments to their charters passed during the early 1860s, the Amoskeag, Nashua, and Great Falls companies were able to obtain much the same benefit offered by a mill act. The amendments gave each company the authority to petition the SJC to have damages for flooded land assessed. The payment of damages would bar any further action for flooding. See Act of 3 July 1861, ch. 2548, [1855–61] N.H. Laws 2483; Act of 8 July 1862, ch. 2683, [1862–6] N.H. Laws 2666; Act of 8 July 1862, ch. 2679, [1862–6] N.H. Laws 2659. Also see *Great Falls* v. *Fernald*, 47 N.H. at 444.

Six justices sat facing the parties in the case of *Winnipiseogee Lake Company* v. *Young*. The chief justice was Samuel Bell. Bell was sixty-one and had just recently been appointed chief justice, although he had been on the court for over ten years. Like his brother James, the Lake Company's agent, he had connections to manufacturing companies and an interest in the productivity of New Hampshire's waters. In the 1830s, he corresponded with Nathan Crosby – an important early figure in the Lake Company venture – about establishing a manufacturing company on the Powow River in East Kingston, New Hampshire (the Powow flows into the Merrimack at Amesbury, Massachusetts). Their letters discuss how to obtain the right to flood the land of neighboring farmers with their proposed dam. Several meadow owners along the river objected to their plans, but Bell had little patience for them. One farmer, whom he characterized as "a man of property & influence," proved especially vehement in his refusal to allow his meadows to be flooded. This man, wrote Bell, "has a prejudice against Factories . . . a violent prejudice . . . growing out of old quarrels about the flowage and is a very selfish man."[63] Bell, however, seemed a ready supporter of the industrial use of water.

Justice George W. Nesmith embraced the factory order with even more enthusiasm. Born in Ireland in 1800, Nesmith received his education at Dartmouth College, a member of the class of 1820 (as was Nathan Crosby). Soon afterward, he began studying law and later helped found the new town of Franklin, New Hampshire. Before his appointment to the superior court, Nesmith had financed a hosiery mill in the town. Together with his cousin John Nesmith – who in the 1840s helped the Lake Company amass land and water rights – he built a two-story mill at the upper dam in Franklin Falls. In 1858, the mill burned down, and the following year Nesmith began his term on the court.[64]

Henry Adams Bellows joined the court that same year. A diligent, committed attorney, Bellows devoted his life to the law in the years before his appointment. Unlike Bell and Nesmith,

[63] S. D. Bell to N. Crosby, 11 July 1835, Bell Family Papers, vol. 2, file 2, N.H. Historical Society, Concord, N.H. For biographical information on Bell, see Charles H. Bell, *The Bench and Bar of New Hampshire* (Boston, 1894), 101–4.

[64] Bell, *Bench and Bar*, 119–21. D. Hamilton Hurd, *History of Merrimack and Belknap Counties, New Hampshire* (Philadelphia, 1885), 36–9, 311. Hurd describes John and George Nesmith as brothers. They were actually third cousins.

he does not appear to have had any major interest in manufacturing. Yet he did have a prior connection to the Lake Company, the plaintiff in the case he would hear in 1860. Two years earlier, Bellows had represented the company when John Coe launched his suit for an injunction.[65]

Three other justices, Charles Doe, Asa Fowler, and Jonathan E. Sargent, rounded out the court. Doe was the court's youngest member at age thirty, a full generation behind his colleagues. Later in the nineteenth century, Doe would go on to become New Hampshire's best-known chief justice, the shining star of the state's judiciary for twenty years. But in 1860, Doe's career as a jurist had just begun.[66] Fowler and Sargent, however, were, aside from Bell, the most experienced justices on the court. Both had been given judicial appointments in 1855, Fowler to the superior court, Sargent to the then existing court of common pleas. (When that court was abolished in 1859, Sargent gained appointment to the SJC.)[67]

These six men were to determine the outcome in the Lake Company's suit. Among them were some who were no doubt sympathetic to the company's cause, men of property with interests in manufacturing and ties to the promoters of the Lake Company. The others, at least on the surface, showed no such prejudices. No evidence suggests that any of the justices had connections to the defendants in the case.

In November 1859, the Lake Company issued a bill for an injunction against the eight people it considered likely to cause it future harm. Josiah French and the Lake Company's attorney, Thomas J. Whipple, wrote the bill. According to one source, Whipple was a leading lawyer in Belknap County during the 1850s, with experience defending corporations.[68] Whipple and French sought to have the court enjoin the defendants from interfering with their dam at Lake Village. They based their plea on one major issue of substance: their legal right to flood land with their dam. That right, they held, derived from two different sources. First, they believed their right to raise the water above Lake Village stemmed from the origi-

---

[65] *Coe* v. *Lake Co.*, 37 N.H. at 254. Also see Bell, *Bench and Bar*, 113–16.

[66] John Phillip Reid, *Chief Justice: The Judicial World of Charles Doe* (Cambridge, Mass., 1967), 77–8. In the important New Hampshire case of *Brown* v. *Collins*, Doe argued in 1873 against a blanket liability for accidents to property as an obstacle to economic progress. See ibid., 134–7.

[67] Bell, *Bench and Bar*, 112, 117.

[68] Ibid., 737.

nal purchase of land and water rights in the area of the dam. More important, they argued that their right to retain the water in Lake Winnipesaukee was founded on "uninterrupted use and enjoyment of that right for more than twenty years." In other words, the company based its right to regulate the water on the doctrine of adverse use. If the defendants owned any land near the lake, the company claimed, that land was not being injured any more than it had been for at least twenty years. They underscored their argument by pointing to the company's excavations between the lake and Paugus Bay; those excavations, they believed, lowered the water level in Lake Winnipesaukee.[69]

To these charges, Young and the other defendants demurred, offering a number of reasons why the court should not grant the company its injunction. To comprehend their objections, especially their response to the issue of adverse use, it is necessary to digress and examine exactly how the Lake Village Dam worked to retain the water above it. Nobody understood the mechanics of the Lake Company's Winnipesaukee water control system better than James Francis.

According to Francis, the structure holding back water at Lake Village was not really a dam. He defined a dam as "a fixed obstruction in a stream over which the whole of the water passes, excepting such as may be used to drive Mills." By this standard, no such dam existed at Lake Village in the 1850s. As he described it, no water was intended to flow over the top of the Lake Village structure. Rather, as the height of the water in Lake Winnipesaukee varied, planks and gates were raised or lowered to control the water. Perhaps this seems a trifling detail. But as Francis realized, the Lake Village "Dam" had some extremely significant legal implications. In its present form, the dam required "constant alteration, and what is much more, no fixed rights are settled by twenty years use." Francis thought this a "fatal objection." Although the structure "has been in use for more than twenty years, and the right is claimed of keeping the water up to a certain point; this claim may be valid, but it is much less definite & unchallengable [*sic*], than the claim to maintain *a dam* at fixed height would be under similar circumstances."[70]

69 *Winnipissiogee Lake Cotton and Woolen Manufacturing Company* v. *George W. Young* (1860), Bill in Chancery, vol. 49:355–60, NHSCN.
70 Memorandum Relating to Dams at Outlets of Reservoirs in New Hampshire, 26 Feb. 1858, A-17, file 82, PLC Papers.

Francis had no legal training. But he did correctly identify the problem the company would have in demonstrating a right to flood land based on adverse use for twenty years. The peculiar structure of the Lake Village Dam – designed with the single-minded intention of storing water – appeared to weaken the company's legal right. Francis's reflections are important because the defendants based their case on precisely what he feared.

Ellery A. Hibbard was the lawyer for the defendants. He was thirty-three when he took the case and had practiced law in Laconia for several years.[71] Hibbard's first two points in the demurrer focused on the company's failure to provide adequate information in its bill. The Lake Company, he argued, had neglected to inform the court of "where, or by what authority, they are in existence, or that they have any existence whatever." Similarly, the bill said nothing about where the company's factories were located.[72] Apart from the posturing present here, these propositions cut to the very essence of the Lake Company's involvement in New Hampshire. The objections questioned the legitimacy of a company that existed to supply water to mills in a neighboring state.

Hibbard also challenged the company's right to legally flood land by virtue of twenty years of adverse use. It did not appear, he argued, "that the use and enjoyment mentioned in said bill . . . has been adverse, and it must appear that it has been uninterrupted and adverse." He further pointed out the bill's failure to make any statement proving "the water in the dam has ever been kept at any definite height at any time."[73] These points followed from the company's methods for regulating the water above Lake Village. The form of the dam at Lake Village primarily served the interests of water storage. It controlled the level of the water in Lake Winnipesaukee to the height necessary to meet the needs of the mills at Lowell and Lawrence. Much as Francis had feared, the defendants attacked the legitimacy of the Lake Company's right to manage the water, its purposeful failure to keep the water at any fixed height. Hibbard was onto the company's scheme.

The court had to decide whether to accept the plaintiff's bill as written or side with the defendants in their demurrer. A

---

[71] Charles W. Vaughan, comp., *The Illustrated Laconian* (Laconia, N.H., 1899), 193.
[72] *Lake Co.* v. *Young*, 40 N.H. at 425.
[73] Ibid., at 425–6.

decision for the plaintiff would clear the way for the company's injunction; otherwise, the bill would be dismissed. In this case, the SJC exercised its authority as a court of equity because the plaintiff in the case, the Lake Company, was seeking an injunction. They had to decide whether the basis of the defendants' demurrer – that the Lake Company had not acquired a prescriptive right to flood the land – was valid. Chief Justice Samuel Bell drafted the court's opinion.

Bell was hardly a novice on the subject of water law. He had written four opinions for the court on the matter during the preceding decade.[74] For the most part, his thoughts meshed with the general direction of water law in mid-nineteenth-century New Hampshire. The opinions paid customary respect to the common-law rule of natural flow, but landed more firmly on the side of reasonable use. In all, they were part of that new corpus of water law emerging after 1850, a flexible body of ideas more consistent with the needs of economic development. Most important are his views on adverse use, the central issue before the court in *Lake Company* v. *Young*. Five years earlier, Bell had written the opinion in *Wallace* v. *Fletcher*. As noted, that case provided an elaborate justification for the doctrine of adverse use. The opinion persuasively disputed the legitimacy of claims founded on prescription, establishing its irrelevance to American courts.

When Bell took up his pen in *Lake Company* v. *Young*, he had to rule on the issue of adverse use. He also had to address the defendants' charge that the company had failed to keep up the water to a definite height. Since water levels fluctuated, he felt it impossible for there to be any continuous use of it. Thus it was also impossible for the company, in staking its claim on adverse use, to be expected to use the water in Lake Winnipesaukee continuously. He compared the situation there with drawing water from a well:

> In the nature of things there cannot ordinarily be any continuous use of such an easement [a right of use over property akin to the Lake Company's right to flood land with its dam]; but if it appeared that a party had claimed the right of drawing water at all times, at his pleasure, and had exercised the right of drawing, as he had occasion, for twenty years, that would establish his right.[75]

---

[74] *Snow* v. *Cowles*, 22 N.H. at 296; *Bassett* v. *Salisbury Mfg. Co.*, 28 N.H. 438 (1854); *Wallace* v. *Fletcher*, 30 N.H. at 434; *Tillotson* v. *Smith*, 32 N.H. at 90.
[75] *Lake Co.* v. *Young*, 40 N.H. at 436.

In other words, Bell believed the Lake Company need not have shown a continuous use of its easement. Rather, *repeated* use was enough to demonstrate the plaintiff's right in this case to use its dam to flood land. The point he made disputed the defendants' claims. At least on this score, the Lake Company's bill, in his opinion, seemed consistent with what the law required.

Bell also struck down the defendants' claim that the company did not properly inform the court of the location of its mills. This argument lacked foundation, Bell said, for it was obvious: "The privilege and mills of the plaintiff, and the dam threatened to be torn down and demolished, are alleged to be at Folsom's Falls, in the towns of Laconia and Gilford, in the county of Belknap."[76] Bell must have known that the Lake Company was managing the water in New Hampshire to benefit the mills downstream in Massachusetts. But the company's larger purposes and behavior were not on trial. Rather, the central legal question remained whether the company's bill for an injunction conformed to what the law required. Thus, Bell upheld the bill's description of the plaintiff's mills. He agreed, however, with the defendants that the bill failed to properly describe the Lake Company as a corporate entity. But because this was a mere technicality, he granted the plaintiff leave to amend its bill.[77] The Lake Company, in short, had won its case.

Lake Company v. *Young* is not a landmark decision. Bell's opinion is not terribly well written; the language is at points vague and imprecise. Yet to the parties involved, the stakes were high, the ruling important. The court had cleared the way for the company to request a permanent injunction, which was subsequently issued. In June 1862, the case appeared before the court when the defendants tried again to have the injunction dismissed. This time George Young chose to write a brief of his own. Young's brief is, in general, a rehash of the same objections presented in 1860, a last-ditch attempt to ward off, however briefly, the inevitable. But it is instructive to look at what Young said about the most substantive issue in the case, the doctrine of adverse use. In his words, "Where he who is in possession of land would avail himself of possession as a bar to the legal title he must show not only an uninterrupted and undisputed possession for twenty years, but a possession ad-

[76] Ibid., at 429.
[77] Ibid., at 427–9.

verse to the title of the legal owner."[78] The language, tortured as it may be, seems remarkably familiar. To claim title to water based on adverse possession, Young pointed out, one had to manage the water in a way contrary to the interests of another owner's property. Interfering with the way someone used their land – by flooding it for example – explicitly notified the landowner that the right to flood was being claimed. If no legal action was taken to prevent the flooding for twenty years, only then could the right to flood by prescription be maintained.

Young's view here was close to Samuel Bell's earlier thoughts on the matter. Apparently the two shared some common ground. It is perhaps ironic, a twist worthy of fiction, that George Young, who struggled for years to deny the Lake Company control over water, defended the doctrine of adverse use. That legal rule proved perfectly consistent with the company's objectives. To obtain control of water and apply it to production on the scale the Boston Associates envisioned presupposed an instrumental, dynamic conception of property. Such an understanding of property involved freeing water from common-law strictures – such as the doctrine of prescription – that interfered with the goal of economic development. As a legal doctrine, adverse use was well tailored to the needs of New England's growing industrial economy.

By the time Young wrote his brief in 1862, the doctrine of adverse use had become an entrenched legal precedent. And Young's argument of course did not refute the legitimacy of this doctrine. Rather, he questioned whether the Lake Company had indeed fulfilled the legal criteria necessary to stake its claim to raise water under that rule. In finding for the company, the court noted that it had repeatedly exercised its right to raise the water, establishing its legal right to do so. The verdict came as no surprise. No conspiracy worked to defraud the defendants, although some of the justices – particularly Bell and Nesmith – were perhaps unduly sympathetic to the Lake Company's cause. But this is not merely a case in which powerful interests combined against their weaker opponents. Rather, there is reason to suspect that a consensus, certain shared values on how water should be used, shaped the outcome. If Young's views are any indication, the defendants may

---

[78] *Winnipisseogee Lake Cotton and Woolen Manufacturing Company* v. *George W. Young* (1862), Defendants' Demurrer, vol. 61:53–6, NHSCN.

have subcribed to some of the same assumptions as the company. The concept of adverse use developed along with other legal rules to make water serve the needs of a developing economy. This vision of water was a thoroughly instrumental one. Young had little to do with how New Hampshire's nineteenth-century water law evolved. But when the case of *Lake Company* v. *Young* went to court, those doctrines posed the framework for the debate. Perhaps a consensus on the need to maximize the productive potential of water helps explain the outcome of the case.[79]

In the spring of 1862, the SJC granted the Lake Company its injunction, ending the chain of events that began with the attack on the dam in 1859.[80] This victory left the company secure for the moment. But the threat mounted by Worster, Young, and the others was not the only challenge the company had to endure. The most important of these other challenges began before the fateful events of early fall 1859. In July 1859, John L. Perley, a prominent Laconia physician and the town's wealthiest citizen, took legal action against the company for flooding his land.[81] Perley claimed the company injured his property when it used flashboards on the Pearson Dam to manage the water in Sanbornton Bay (Lake Winnisquam). In response, the Lake Company obtained a temporary injunction preventing Perley's case from going any further. The company then brought a countersuit involving issues that they claimed needed consideration before Perley's case could be tried fairly.[82]

The facts of the case were long and complicated. In 1852, the Lake Company and John Perley signed a deed giving the company the right to flood Perley's land. A provision in the deed prevented the company from raising the water to a point more than twenty inches above Eager Rock on the east side of Sanbornton Bay. Like the dam at Lake Village, the Pearson Dam

---

[79] My thoughts here were influenced by Elizabeth Fox-Genovese and Eugene D. Genovese, *Fruits of Merchant Capital: Slavery and Bourgeois Property in the Rise and Expansion of Capitalism* (New York, 1983), 345–6, and by Tushnet, "Marxist Analysis of American Law," 100–101, 106–7.

[80] *Lake Co.* v. *Young* (1862), vol. 61:71–4, NHSCN.

[81] For biographical information on Perley, see Vaughan, *Illustrated Laconian*, 163–5. Knowledge that Perley was the wealthiest citizen in Laconia is from Manuscript Schedules, U.S. Census of Population, Laconia, N.H., 1860.

[82] J. French to J. T. Stevenson, 18 Mar. 1864, item 204, EC Papers.

also used flashboards to control the water above. Although the rock sat several miles from the dam, it was near enough to Perley's residence to let him check on the company. Perley and James Bell, who represented the Lake Company at the time, believed a depth of water of twenty inches on Eager Rock equalled fifteen inches of water on the Pearson Dam. The deed, however, did not explicitly specify this.[83]

Unfortunately, the Lake Company soon found that twenty inches of water on the rock corresponded to less than fifteen inches of water at the dam, a fact depriving it of the bay's full potential to store water. Apparently, Bell and Perley had an unwritten agreement on how the water should be regulated, an agreement satisfying all concerned until 1859. According to Josiah French, who was the agent when Perley brought his action, matters changed when Bell passed away:

> The deed as Mr. Bell took it was all very well as long as Mr. Bell lived, for he could explain the agreement and the deed, and Doct. Perley never expressed any doubts as to the Company's rights as they were exercised under the deed, till the death of Mr. Bell, but in July 1859 about two years after Mr. Bell's death Dr. Perley brought a suit for flowage alleging that the company kept the water in Sanbornton Bay higher than they had a right to by virtue of the deed which he gave.[84]

Why Perley waited seven years to bring his case to court is not clear, but he may have been motivated by vengeance. Clearly the agreement with the company had satisfied Perley for some time. The Lake Company after all had paid him a considerable sum of money ($15,000) for the right to flood his land. Why did Perley change his mind?

As part of the agreement signed in 1853, the Lake Company pledged to enlarge an opening in the Pearson Dam, called a wasteway, to sixty feet. The wasteway let the company release surplus water (that is, any above the fifteen inches on the dam to which it was entitled), when the waters of the bay were swollen. But the Lake Company did not completely live up to its word. When finished, the wasteway measured but fifty-eight feet in width, giving Perley less than the agreed upon security during the high-water season. By French's own admission, the company also failed to keep the water on the dam below fifteen inches.[85]

[83] *Lake Co.* v. *Perley,* 46 N.H. at 83–99.
[84] J. French to J. T. Stevenson, 18 Mar. 1864.
[85] Ibid.

In 1865, the case came before the SJC. Again the court applied the principles of equity to the case. The company asked the court to reform the original deed to account for its right to keep fifteen inches of water on the dam. Perley responded that the agreement allowed no more than twenty inches of water on Eager Rock regardless of the level at the dam. His contention was not entirely made in good faith. The Lake Company had not paid Perley $15,000 to flood a rock; it paid him for the privilege of using fifteen-inch flashboards to store water above the dam. Each side brought forth armies of witnesses and a voluminous amount of evidence. The court considered the matter and ruled for the Lake Company, ordering the deed reformed.[86]

This left Perley free to pursue the company for flooding his land. To the Lake Company, the stakes were meager, of little consequence given the outcome of the earlier case. True, the company had failed to enlarge the wasteway to sixty feet. But Josiah French believed the damages awarded would be small. In September 1865, French was able to settle with Perley out of court for a mere hundred dollars.[87] Once again, the Lake Company had prevailed. It had now triumphed over a range of opponents – wealthy and poor, farmer, professional, mechanic, and laborer. In the two decades since the Boston Associates first sought control of the waters of New Hampshire, there were precious few years of peace. Yet somehow the embattled company had wriggled free of its opponents and seen its way to victory.

In 1868, the New Hampshire legislature passed that state's first mill act in three-quarters of a century. The previous year, the Committee on Manufactures of the House of Representatives had advised the legislature to pass a mill act to encourage manufacturing. In their opinion, the "foolish jealousies" opposed to industry had withered; New Hampshire, it appeared, seemed on the brink of a more enlightened age. As they wrote:

> No one that has mingled in the society of New Hampshire men of all political parties, can resist the conclusion that the intelligent farmers, mechanics, manufacturers, and all other business men, not only consent to, but demand the enactment of a general law, which will

---

[86] *Lake Co. v. Perley*, 46 N.H. at 83; also see *Winnipissiogee Lake Cotton and Woolen Man'g Co. v. John L. Perley* (1865), Depositions, N.H. Reservoirs Pamphlet Box, Special Collections, University of Lowell, for the evidence marshaled by both sides.

[87] J. French to J. Francis, 28 Sept. 1865, A-39, file 215, PLC Papers.

effectually protect the interest of the landholder, and, at the same time, secure stability and protection to the capital seeking investments within our borders.[88]

The mill act removed the threat dam breaking posed to manufacturing companies. It provided for a committee appointed by the SJC that would investigate land injured by the raising of a milldam on a non-navigable stream. Should the committee discover a dam to "be of public use or benefit to the people of this state," it was to estimate the damages to be paid.[89] The law undermined the protection afforded land from flooding, long a sacrosanct aspect of the state's water law. The act thus further freed New Hampshire from the restraints of the common law. It made the state safer for industrial capital by putting the law more firmly on the side of dam owners.

To New Hampshire's waterpower companies, the 1868 mill act was a victory soured by the long delay in passage. In the time it took to pass the law, a good deal of water had flowed through their mills. But the state's failure to enact such legislation did not deter the Lake Company from carrying out its plans. Even without a mill act, the company managed to assume control of an immense share of the state's water wealth. To what did the company owe its success?

While New Hampshire legislators were reluctant to pass a mill act, their counterparts on the judiciary moved to shape the state's water law. Especially after midcentury, the law increasingly accounted for water's productive value. In legitimizing the doctrines of reasonable and adverse use, New Hampshire jurists bolstered the legal foundation on which manufacturing interests based their claims to water. But this alone does not account for the Lake Company's stunning success. The powerful Lake Company also had some friends in high places, and surely the role of a sympathetic judiciary should not be ruled out. How striking that in its five cases before the state's highest court, the Lake Company never suffered a defeat.

The Amoskeag Company, however, was much less successful, at least in the cases it tried before the SJC. In *Eastman* v. *Amoskeag Manufacturing Company* (1862) and *Amoskeag Manufacturing Company* v. *Goodale* (1865), the company lost when strict rules regard-

---

[88] *Journal of the House of Representatives of the State of New Hampshire* (Manchester, N.H., 1867), 2 July 1867, 367.

[89] Act of 3 July 1868, ch. 20, [1868] N.H. Laws 152.

ing liability for flooding land were applied.[90] In the latter case, the court overturned on appeal a lower court verdict in favor of the company. The company brought action for trespass against the defendant, Edwin E. Goodale, for destroying the flashboards on its dam.[91] The law, in the court's opinion, was perfectly clear: A riparian landowner whose dam flowed water back onto the land of another "perceptibly higher than its natural level" was liable for nominal damages. In other words, whether or not the dam caused any *actual* injury, the mere change in the water level exposed the dam's owners to the payment of damages and possible legal action. According to the opinion, the jury at the lower court's trial had been misinstructed. If the dam had caused no "appreciable or substantial damage" to the defendant's land, the jury instructions read, then he had no right to enter the company's property to remove the nuisance. Overruling the earlier verdict, the SJC upheld the defendant's legal right to abate the nuisance, even though the dam caused only nominal damage to his land.[92]

In 1865 when the case was appealed, New Hampshire had not yet passed the mill act. To gain a legal right to flood land, mill owners had to negotiate such rights on an ad hoc basis. With no formal procedure for judging damages caused by dams – a situation redressed by the passage of the mill act in 1868 – waterpower companies at times found themselves involved in litigation with no guarantee of victory. Their defeats in cases like *Amoskeag* v. *Goodale* were powerful reminders of the potential hostility of New Hampshire's legal climate.

Generally, however, New Hampshire's water law favored large, industrial users over other competing claimants. The Lake Company's victory against George Young and the other defendants is an example. The legal rule of adverse use encouraged maximum development of water and gave the company an edge in the case. But the law of water by itself did not insure a victory for the company. Although the Lake Company and the defendants opposed one another in court, they shared cer-

---

[90] *Eastman* v. *Amoskeag*, 44 N.H. at 143; *Amoskeag* v. *Goodale*, 46 N.H. at 53. In *Eastman* v. *Amoskeag*, the court ruled that liability for flooding cannot be exonerated by the conveyance of land. It further said that an act of the legislature chartering a corporation does not protect the corporation, in this case the AMC, from the liability for flooding land. See *Eastman* v. *Amoskeag*, 44 N.H. at 143, 159–60.

[91] Goodale, along with Worster and others, had also attacked the dam at Amoskeag Falls in the summer of 1859. See Journal of Ezekiel Straw, 27 July 1859.

[92] *Amoskeag* v. *Goodale*, 46 N.H. at 53–4, 56–7.

tain assumptions that helped the company to succeed. In this case, the defendants failed to question the instrumental conceptions underlying the law of water. Admittedly, this was not the defendants' central task in the case. Yet few of the Lake Company's opponents seemed eager to challenge the company's right to use water productively for profit. Rather, they objected to the particular course taken by the company, to its greed in using the water to aid factories in Massachusetts. Into whose pocket did the productive value of water fall? Here was the main source of discontent. There were perhaps a few followers of Henry Thoreau left who opposed any productive use of water. But for the most part, there seems to have been a consensus among those who fought for control of New Hampshire's water that the flow should be valued and employed for its economic potential. Given this view of water and how it ought to be used, the Lake Company's victories are not surprising. Who knew better how to manipulate water in the name of production?

From their rather humble beginning at Chelmsford, Massachusetts, the Boston Associates set off to master the waters of the Merrimack valley. In the years from 1822 to 1845, they built a string of mill towns and engineered intricate systems for distributing water to produce textiles. When the river proved unable to meet their ambitions for production, they sought a broader control and went determinedly after the New Hampshire lakes. They purchased the Lake Company, supplied it with ten times the capital stock, and then used it to achieve an impressive amount of control over New Hampshire's vast waters. The company ingeniously put together a well-designed mosaic of land and water rights, constructing an elaborate water control infrastructure set up to store water for Massachusetts industry. It fought tenaciously, at times insidiously, to keep control of the water in the face of opposition. And in the end, the company consolidated its claims by skillfully succeeding in New Hampshire's highest court.

As this process unfolded across two New England states, it produced as many costs as it did fabric and profits. Placing the valley's waters under industrial control meant wresting it from those with a different agenda – farmers, petty producers, and others. The latter were the losers. At the very least, they lost control over the water as a resource. For some, this may have meant the loss of livelihood, of control over their economic destiny. For others, it may have simply produced intense resentment. There were, however, other costs as well. Watersheds are delicate and at times easily damaged by human activity. Environmental costs resulted from the rush to make water serve the needs of industry – costs not easily made up.

6

DEPLETED WATERS

In the summer of 1867, Benjamin Coolidge, an assistant engineer at Lawrence, accompanied James Francis and Charles Storrow to New Hampshire to visit the Lake Company. Although they journeyed on business, not pleasure, Coolidge remarked on the natural world around him. A steamer brought them across Lake Winnipesaukee, and Coolidge peered down into the clear, slightly green water. He did not mention whether he saw any fish. But he did spot Mount Washington and "the foliage in the distance looking as even as velvet, and of a very rich green." Looking up he caught sight of three loons looming overhead, their white breasts beaming down at him. They stopped at Centre Harbor to rest, and after taking tea, Coolidge and Storrow strolled behind their hotel. From there, they had a splendid view across Lake Winnipesaukee. "We are evidently breathing mountain air," wrote Coolidge, "and exhilarating air it is." He breathed deeply, taking in the air, the scenery, perhaps knowing – tucked away in the back of his mind – that he would soon need to return to the urban sprawl that was Lawrence, Massachusetts.[1]

Benjamin Coolidge still found much natural beauty in the Merrimack valley. Yet the environment he viewed in 1867 bore the unmistakable imprint of industrial transformation. The New England he and his colleagues traveled through, the land whose waters they had played no small part in altering, encompassed a visibly different ecology. True, the loons, the mountain scenery, and serene waters were still there. But one important part of nature's complex mosaic – the fish – was missing. While Coolidge was off enjoying the region's natural beauty, others were describing a grimmer reality, a waterscape that was suffer-

---

[1] Benjamin Coolidge Red Letter Notebooks, 31 July 1867, bk. "M," item 69; also see C. Storrow to B. Coolidge, 23 July 1867, item 24, EC Papers, MATH.

ing. According to an 1867 assessment, New England's salmon, shad, alewives, and other species of freshwater fish had "a half century ago, furnished abundant and wholesome food to the people; but, by the erection of impassable dams, and needless pollution of ponds and rivers, and by reckless fishing . . . our streams and lakes have been pretty much depopulated."[2]

Once a flourishing, vital resource, the Merrimack River fisheries were essentially destroyed by the middle of the nineteenth century. A number of factors contributed to their decline, but the completion of the Lawrence Dam in 1848 was decisive. A massive structure over thirty feet in height, the dam blocked migrating fish – salmon, shad, and alewives – from passing upstream to spawn. There were other barriers preventing these species of fish from traveling freely throughout the river, but the dams constructed by the Waltham–Lowell mills were the most formidable. They became an obvious focus for attack as New Englanders wrestled with the environmental consequences of industrialization.

River fisheries are, ecologically speaking, a delicate and vulnerable resource. As predators deplete a river's stock of fish, it can be renewed only if enough fish remain for breeding. Obstructing a river interferes with this process by cutting fish off from their spawning grounds. In addition, more subtle factors may be at work. The physical quality of river water – its temperature, velocity, volume, and chemical balance – affects the ability of particular species to reproduce.[3] Temperatures above twenty-seven degrees centigrade, for example, tend to kill the fish eggs of Atlantic salmon. Higher water temperatures also cause young fish to develop more rapidly and return to the sea earlier. And when there is little rainfall and the volume of water discharged by a river remains low throughout the year, research suggests a tendency toward a smaller surviving population of fish.[4]

Human beings pose an even greater threat to river fisheries. None of nature's predators has the sharp capacity for reasoned

---

[2] New England Commissioners of River Fisheries, Circular, 26 Feb. 1867, vol. 20, file 103, PLC Papers, BL.

[3] A fine discussion of fish and river ecology can be found in V. M. Brown, "Fishes," in *River Ecology*, Studies in Ecology, vol. 2, ed. B. A. Whitton (Berkeley, 1975), 199–229. A good general text on fishery biology is Carl E. Bond, *Biology of Fishes* (Philadelphia, 1979).

[4] Kenneth D. Carlander, *Handbook of Freshwater Fishery Biology*, 2 vols. (Ames, Iowa, 1969), 1:204–7.

thought that make human beings so potentially harmful to other species. Fisheries are especially vulnerable because, by their very nature, they are a common resource. They are common in the sense that it is difficult for any one person or group to exclude others from competing for the fish. This fact can easily result in depletion, for every fisherman knows that a fish not caught is potential bounty for others. Fisheries are often a perfect target for the greedy, a dilemma that can ultimately lead to overfishing and eventual ruin.[5]

Different cultures tend to natural resources in their own unique ways. A culture with a strong affinity for private property and markets finds the regulation of fisheries especially problematic. And where an existing system of exchange reduces fish to a cash-valued commodity, overfishing may often result. The logic of such an economic system, a system driven by profits and increased production, can at times compel reckless exploitation. For capitalist cultures then, fisheries raise tough, intractable problems that resist even the best attempts at governmental intervention. From almost the first years of settlement, New Englanders relentlessly passed fishery legislation. But the effective enforcement of those laws turned out to be no small task. Government regulation raised its own concerns, questions about whether the public's access to fisheries should be suspended in favor of economic growth, of the propriety of sacrificing public right for private benefit.

The pages that follow adopt a rather wide optic. They survey the ecological history of the Merrimack valley fisheries, their decline, and the attempt to restore them later in the nineteenth century. As industrial development intruded on the valley's waters, a sweeping ecological transformation occurred, paralleling the changes, it might be added, happening in work, family, gender, and culture. Our task will be to understand how the Waltham–Lowell system and other factors figured in the ecological decline of the valley's once thriving fisheries. In the 1860s, the devastation of the region's fish stocks inspired an effort to revitalize this now depleted resource. This too is part of our story, for it suggests the options open to a region that had thoroughly industrialized its waterscape.

* * *

[5] See p. 14, n. 51. For a historical perspective on common resources, see Arthur F. McEvoy, *The Fisherman's Problem: Ecology and Law in the California Fisheries, 1850–1980* (Cambridge, 1986).

Matthew Patten, an eighteenth-century farmer who lived in Bedford, New Hampshire, liked to fish for shad in the Merrimack River. In some respects, Patten was an ordinary New England farmer, engaged in the daily and seasonal pursuits of agricultural life – planting, harvesting, hunting, fishing, and church going. Yet in at least one respect, Patten was rather extraordinary. As anyone who has tried it knows, to keep a diary for over thirty years takes tremendous diligence. The thoughts he recorded there are scarcely profound, focused as they are on his daily routine. But taken as a whole, the document is a remarkable statement of how a colonial farmer spent time coping with the demands of daily existence, the perhaps mundane, but no less essential, routines of life. Among other things, the entries show that between April and the end of June, Patten devoted a large share of his time to fishing (and hunting). By the time he died in 1795 – keeling over in a field – he had lived through a period in the Merrimack River's ecological history that his nineteenth-century descendants would never know firsthand.[6]

Patten's fishing picked up significantly during the spring when the Merrimack's most important species of fish entered the river to spawn. At Amoskeag Falls, where he ventured primarily to fish, salmon, alewives, lamprey eels, and shad were common. Atlantic salmon were the river's prize fish, but the other species were far more abundant here. The most important period of migration was from April through June. During this time, shad, alewives, and salmon were all entering the river in search of gravel beds along the shore or river bottom to deposit their spawn. Although lamprey eels also ascended the river in the spring, it was during their fall descent downstream that fishermen had the most success in trapping them.[7] These species of fish generally spent most of their lives at sea. But when they migrated to the freshwater Merrimack to breed, they inadvertently gave fishermen ample opportunity to capture them.

[6] Matthew Patten, *The Diary of Matthew Patten of Bedford, N.H.* (Concord, N.H., 1903). The diary is analyzed in Thomas C. Thompson, "The Life Course and Labor of a Colonial Farmer," *Historical New Hampshire* 40 (1985): 146.

[7] On the Amoskeag fishery, see George E. Burnham, "Amoskeag's Old Fishing Rocks," *Manchester Historic Association Collections* 4 (1908): 60–7; C. E. Potter, *The History of Manchester, Formerly Derryfield, in New-Hampshire* (Manchester, N.H., 1856), 641–51; Myron Gordon and Philip M. Marston, "Early Fishing Along the Merrimack," *New England Naturalist*, Sept. 1940, 3–6. Detailed information on the biology of fishes is from Carlander, *Freshwater Fishery Biology*, 1:70–82, 196–207.

White European settlers – people like Patten who came to New England from Ireland in 1728 – were not, of course, the first to discover the river's fisheries. Archaeologists studying the early human cultures of the Merrimack valley have made some suggestive findings. They have found that the majority of occupied sites lay within less than two miles of falls or rapids on either the Merrimack or its tributaries. The sites were chosen, it is believed, for the convenience they offered as transportation points and/or for their advantage as fishing sites. Further evidence indicates a preference for sites positioned just downstream from falls or rapids, better locations for catching migrating fish.[8]

Whatever the spring fish runs at Amoskeag Falls meant for the Indians, they proved especially attractive to the white settlers who followed. Using spears, seines, and scoop nets, New Englanders fished from the rocks at Amoskeag Falls, laying claim to close to twenty-five separate fishing places.[9] In the space of one mile, the Merrimack descends fifty-four feet at this point. The force of the current caused fish to seek shelter behind the large rocks at the falls, behavior that made them easy prey for fishermen.[10] Before the powerful waters enticed Boston capitalists to invest their money here, Amoskeag Falls remained the region's best-known spring fishery. During the eighteenth century, Patten and several others constructed a seine and staked claim to a spot in the river.[11] Their site had its moments, both more and less prosperous. In the first week of June 1772, to take one example, Patten ventured to the falls for six straight days. He kept careful track of his yield, which totaled 551 shad for the week.[12] There were other times when Patten was down on his luck, but the Amoskeag fishery regularly drew him back to the Merrimack's shore.

Much of the fish taken in the valley wound up on tables nearby, usually smoked or salted down and stored in cellars to be eaten during the long winter months. Otherwise, the fish

---

[8] Victoria Bunker Kenyon and Patricia F. McDowell, "Environmental Setting of Merrimack River Valley Prehistoric Sites," *Man in the Northeast* 25 (1983): 13, 15–18. Also see Dena F. Dincauze, *The Neville Site: 8,000 Years at Amoskeag, Manchester, New Hampshire* (Cambridge, Mass., 1976), 1, 8–9, 133.

[9] Burnham, "Amoskeag's Old Fishing Rocks," 61.

[10] For a discussion of the behavior of fishes in rivers, see H. B. N. Hynes, *The Ecology of Running Waters* (Toronto, 1970), 309.

[11] Thompson, "Colonial Farmer," 153.

[12] Patten, *Diary*, 284.

reached nearby urban markets. Fish from Pawtucket Falls in Lowell were commonly brought to Boston for sale. "The quantity of salmon, shad and alewives caught in Chelmsford [Pawtucket Falls] annually," wrote one observer in 1820, "may be computed at about 25 hundred barrel, besides a large quantity of other fish of less value."[13] Here, much as at Amoskeag, local residents competed for the best fishing places, with forty-two proprietors at one time owning rights.[14] Throughout the valley, at falls on the Merrimack River and at many points on the river's tributaries, fishing played an important role in the regional economy through the early nineteenth century. On the Winnipesaukee River, weirs were stuffed with grass and weeds to trap eels during the fall, while spring brought shad up the river in great numbers.[15] One hundred river miles downstream at Pentuckett Falls in Haverhill, salmon and alewives were caught and cured.[16] And in between, along the Merrimack's tributaries, fish provided a source of dietary protein, easily available during the spring season.[17] Fish were, in short, a critical link in the Merrimack valley's food chain.

The task of managing the Merrimack valley's fisheries proved to be no easy matter. Between 1764 and 1820, New Hampshire passed fourteen legislative acts concerning fishing in the Merrimack River.[18] The enactments are a jumble of regulations, strictures, and penalties. They limited the number of days on which fish could be legally caught, regulated the size of seines, and imposed fines for those who obstructed the river during the

13 Quoted in Frederick W. Coburn, *History of Lowell and Its People*, 3 vols. (New York, 1920), 1:62–3.
14 Silas R. Coburn, *History of Dracut, Massachusetts* (Lowell, 1922), 289.
15 M. T. Runnels, *History of Sanbornton, New Hampshire*, 2 vols (Boston, 1882), 1:277–80.
16 George W. Chase, *The History of Haverhill, Massachusetts* (Haverhill, 1861), 120.
17 Local historians often referred to the valley's fisheries. See, e.g., Peter P. Woodbury, Thomas Savage, and William Patten, *History of Bedford, New-Hampshire* (Boston, 1851), 204–5; Kimball Webster, *History of Hudson, N.H.* (Manchester, N.H., 1913), 23.
18 Act of 8 May 1764, ch. 2, [1745–74] N.H. Province Laws 340; Act of 11 May 1767, ch. 1, [1745–74] N.H. Province Laws 407; Act of 28 May 1773, ch. 6, [1745–74] N.H. Province Laws 603; Act of 9 Apr. 1784, ch. 2, [1776–84] N.H. Laws 547; Act of 26 June 1786, ch. 26, [1784–92] N.H. Laws 186; Act of 6 Feb. 1789, ch. 56, [1784–92] N.H. Laws 412; Act of 18 June 1790, ch. 18, [1784–92] N.H. Laws 527; Act of 12 Jan. 1795, ch. 19, [1792–1801] N.H. Laws 221; Act of 20 Dec. 1797, ch. 50, [1792–1801] N.H. Laws 476; Act of 27 Dec. 1798, ch. 37, [1792–1801] N.H. Laws 544; Act of 20 June 1811, ch. 43, [1811–20] N.H. Laws 45; Act of 23 June 1818, ch. 26, [1811–20] N.H. Laws 685; Act of 1 July 1819, ch. 70, [1811–20] N.H. Laws 822; Act of 16 Dec. 1820, ch. 20, [1811–20] N.H. Laws 937.

seasonal fish runs. Much the same experience prevailed in Massachusetts. That state passed seventeen such acts for the Merrimack in the years from 1783 to 1820.[19] A sense of desperation fills the enactments, some of which openly admit the failure of prior legislation. But whether acknowledged or not, the sheer bulk of the legislation, the continued passage, repeal, and reenactment, suggests how trying it could be to regulate the fisheries.

The reasons for failure are not too hard to find. Although fish wardens were elected by towns to oversee compliance with the legislation, enforcement was virtually impossible. One local historian wrote that fishermen at the Amoskeag fishery used various means, including bribery, to evade fish wardens. "A history of unlawful fishery at the Falls," he noted, "would be more voluminous than interesting."[20] An observer of the fishery at Pawtucket Falls recalled the following episode:

> The law allowed us to fish two days each week in the Concord [River], and three in the Merrimack. . . . The fish wardens were the state police.
> The Dracut folks fished in the pond at the foot of Pawtucket Falls. They would set their nets on the forbidden days. On one occasion the fish wardens from Billerica came, took and carried off their nets. The wardens, when they returned to Billerica, spread the nets on the grass to dry. The next night the fishermen, in a wagon with a span of horses, drove to Billerica, gathered up the nets, brought them back, and reset them in the pond.[21]

The law could only stipulate and penalize. It could not legislate restraint, and for this reason the Merrimack River fisheries remained prone to overindulgence. But the resulting ecological consequences are obscure and difficult to comprehend. Without more complete evidence, the question of how much pressure overfishing placed on eighteenth-century fish populations remains mostly speculation.

Dams built across the river posed a much clearer threat to the river's fisheries. Before the construction of dams, natural barriers such as waterfalls were the major obstacles in the path of migrating fish. Some fish jumped over waterfalls, especially Atlantic salmon, which can clear a distance of eleven feet. The

---

[19] Oscar Handlin and Mary Flug Handlin, *Commonwealth: A Study of the Role of Government in the American Economy* (1947; rev. ed., Cambridge, Mass., 1969), 72.

[20] Potter, *History of Manchester*, 651.

[21] Quoted in Alfred Gilman, "Lowell," in *History of Middlesex County, Massachusetts*, ed. Samuel Adams Drake, 2 vols. (Boston, 1880), 2:56.

construction of dams across the Merrimack, however, interfered with migratory species of fish in several ways. Such
obstructions blocked fish that were less powerful swimmers
than salmon. In addition, dams were troublesome for fish, such
as salmon smolts (a young salmon) and lampreys, passing
downstream. Fish swimming downstream tended to follow the
contour of the streambed and were prone to being caught on
the bottom of the upper side of a dam.[22] Finally, dams affected
the river's streamflow. In general, fish are likely to migrate
when stimulated by rising water. Holding back the water and
reducing the discharge downstream may thus have hindered
the upstream migration of fish.[23]

By the nineteenth century, the progressive obstruction of the
Merrimack River had interfered with migratory species of fish.
The completion by 1814 of a system of canals linking Boston
and Concord, New Hampshire, involved the construction of
dams across the Merrimack River. The total number of dams is
unclear, but the Wicasee Canal, Cromwell's Canal, Union Locks
and Canals, and Amoskeag Canal all had at least one or more,
making fish passage problematic.[24] Further, a new dam built in
Bristol, New Hampshire, in 1820 blocked salmon from reaching a spawning area.[25] "Until within a few years," wrote two
observers in 1824, "salmon and shad were caught near this
village [Piscataquog, which was downstream from Amoskeag
Falls] in sufficient quantities to supply the surrounding inhabitants; but since the erection of locks and dams on the Merrimack, very few have been caught."[26]

The rise of the Waltham–Lowell system, specifically the construction of dams at Lowell and Manchester, extended the process of destruction. Although the river remained clear between

[22] Hynes, *Ecology of Running Waters*, 353–4.
[23] Ibid., 353; James D. McCleave et al., eds., *Mechanisms of Migration in Fishes* (New York, 1984), 325.
[24] On the Merrimack River canals, see Christopher Roberts, *The Middlesex Canal: 1793–1860* (Cambridge, Mass., 1938), 128–33. For a discussion of the ecological effects of dams, see J. C. Fraser, "Regulated Discharge and the Stream Environ-ment," in *River Ecology and Man*, ed. Ray T. Oglesby, Clarence A. Carlson, and James A. McCann (New York, 1972), 263–85.
[25] Lawrence Stolte, *The Forgotten Salmon of the Merrimack* (Washington, D.C., 1981), 4; P. J. Dalley, "Fish Passage Facilities Design Parameters for Merrimack River Dams: Essex Dam Lawrence, Massachusetts and Pawtucket Dam Lowell, Massachusetts" (ms. pre-pared for U.S. Fish and Wildlife Service, Boston, 1975), 8.
[26] A. Foster and P. P. Woodbury, "A Topographical and Historical Sketch of Bedford, in the County of Hillsborough," *Collections of the New-Hampshire Historical Society* 1 (1824): 296.

Figure 6.1. A view of the Lawrence Dam. (Courtesy of the Museum of American Textile History.)

Lowell and the mouth at Newburyport, the upstream spawning grounds were cut off, a change with more impact on salmon than on other species. Shad, although capable of spawning further upstream, typically do so just above brackish water. But salmon seek upstream waters where the swift current maintains the necessary high levels of dissolved oxygen around their eggs. By the middle of the nineteenth century, the Merrimack River began a new phase in its ecological history. Its once flourishing fisheries were teetering on the edge of destruction.

The completion of the formidable Lawrence Dam delivered the final blow to the valley's fisheries.[27] Thirty-two feet high and 1,600 feet across, the dam, whatever its technical merits, had ecological implications that ramified throughout the Merrimack valley (Figure 6.1). The original 1845 act of incorporation contained a provision requiring the builder, the Essex Company, to provide a suitable fishway to allow passage to migrating species.[28] Although fishways built in the early nineteenth century were crude affairs, inclined wooden sluices often poorly placed, they still reduced the amount of water

[27] See Stolte, *Forgotten Salmon of the Merrimack,* 5; Dalley, "Fish Passage Facilities," 8.
[28] Act 20 Mar. 1845, ch. 163, [1843–5] Mass. Acts & Resolves 483, §§ 5 and 7.

available for power.[29] A fishway thus threatened plans for the technological domination of water at Lawrence. Charles Storrow, the mastermind of Lawrence's waterpower system, consulted Benjamin Robbins Curtis, a prominent Massachusetts lawyer, for legal advice on the fishway. The news he gave could not have heartened Storrow. "I am sorry to be obliged to say," wrote Curtis in 1847, "that both individuals and the public have so strong a hold on you." In Curtis's opinion, the company had to build a fishway that satisfied the expectations of county officials. Not to do so would leave it open to litigation and, worse, the revocation of its charter.[30]

With the dam nearing completion in 1847, county officials ordered the Essex Company to build a wooden fishway at least thirty feet wide and fifty feet long. The fishway was to run – rising no more than one foot in height for every four feet in length – from the streambed below the dam to within two feet of the dam's stone crest.[31] When completed, the passage did little to help fish negotiate this towering obstruction. The fishway evidently failed, probably because it was poorly placed for attracting fish.[32] A compromise solution to the fishway problem was worked out in 1848 through an amendment to the Essex Company's charter. The Massachusetts legislature authorized the company to increase its capitalization by $500,000, provided it assumed liability for damages to fishing rights above the dam.[33] For a sum of over $26,000, the company settled the bulk of the claims, releasing it from the obligation of constructing an effective fishway.[34]

Not everyone seemed satisfied with the compromise. In 1847, before completion of the Lawrence Dam, William McFarlin purportedly caught nine hundred shad in one day at his fishing privilege upstream at Pawtucket Falls.[35] Two years later,

[29] On the failure of fishways built before 1950 to pass significant numbers of fish, see Dalley, "Fish Passage Facilities," 5.

[30] B. R. Curtis to C. Storrow, 11 Aug. 1847, "Record of Legal Opinions," item 7, EC Papers.

[31] The plans ordered by the county commissioners of Essex County are summarized in Mass. Legislature, Joint Special Committee on the Obstruction of the Connecticut, Merrimack, and Saco Rivers, "Report" (Apr. 1865), S. Doc. 183, 15. On the Essex Company's compliance with the commissioners' instructions, see Petition of the Essex Company to the Mass. Legislature (Jan. 1857), Mass. S. Doc. 15, 4.

[32] On fishways and why they are sometimes ineffective, see Geoffrey E. Petts, *Impounded Rivers: Perspectives for Ecological Management* (New York, 1984), 231–3.

[33] Act of 9 May 1848, ch. 295, [1846–8] Mass. Acts & Resolves 773.

[34] Petition of the Essex Company to the Mass. Legislature, 6.

[35] This fact was mentioned in a later legislative investigation. See Mass. Legislature,

McFarlin charged that the Essex Company had not "made & maintained suitable and reasonable Fishways but have so constructed their dam as to entirely stop and impede the passage of Fish up & down said [Merrimack] river."[36] McFarlin later petitioned county officials to investigate. A sheriff's jury appointed to decide the matter confirmed McFarlin's claim and awarded him damages. That verdict was then accepted by a court of common pleas. The Essex Company appealed the lower court's decision before the Massachusetts SJC in 1852.[37]

The case of *McFarlin v. Essex Company* turned on one issue: whether McFarlin had properly established his right to the fishing privilege. That right had a rather complicated history, having been established in 1814 and passed back and forth over the decades. McFarlin rested his claim to the fishing place at Pawtucket Falls on adverse use. Delivering the court's opinion, the well-known Massachusetts jurist Lemuel Shaw wrote: "It is now perfectly well established as the law of this commonwealth, that in all waters not navigable in the common-law sense of the term . . . the right of fishery is in the owner of the soil upon which it is carried on."[38] In other words, an exclusive right to fish in non-navigable rivers stemmed from ownership of the riparian land. Navigable rivers and streams, conversely, were considered public, and no exclusive right to fish derived from landownership alone. The court found that McFarlin could not prove ownership of the riparian land. Nor could he establish any prescriptive claim to the fishery without maintaining a prescriptive right to the land. Since McFarlin failed to show adverse, uninterrupted use of the fishery for at least twenty years, the lower court's verdict was set aside.[39]

Although the Essex Company won this case, the construction of the Lawrence Dam remained a source of conflict. Apart from destroying fishing rights, the dam also overflowed land and allegedly damaged at least one mill privilege. In *Hazen v. Essex Company* (1853), the owner of a mill upstream from the dam brought an action for the flooding of his waterwheels before the Massachusetts SJC.[40] In defense, the company

Commission on Obstructions to the Passage of Fish in the Connecticut and Merrimack Rivers, "Report of the Commissioners" (Jan. 1866), S. Doc. 8, 39 (hereafter cited as Mass. Report on Obstructions).

[36] Petition of William McFarlin, 10 July 1849, item 14, EC Papers.

[37] *McFarlin v. Essex Company,* 10 Cush. 304 (Mass., 1852).

[38] Ibid., 309.

[39] Ibid., 312.

[40] *Hazen v. Essex Co.,* 12 Cush. 475 (Mass., 1853).

claimed its act of incorporation sanctioned its right to damage private property by eminent domain. During the nineteenth century, the doctrine of eminent domain allowed the appropriation of private property for public purposes, largely for roads, canals, and railroads.[41] In this case the question, as Chief Justice Shaw explained, was whether the legislature properly granted the Essex Company the right of eminent domain, enabling it to take private property for public use. Was the company really taking private property for the benefit of the public? The court was unequivocal:

> The establishment of a great mill-power for manufacturing purposes, as an object of great public interest, especially since manufacturing has come to be one of the great public industrial pursuits of the commonwealth, seems to have been regarded by the legislature and sanctioned by the jurisprudence of the commonwealth, and, in our judgment, rightly so, in determining what is a public use, justifying the exercise of the right of eminent domain.[42]

Ordinarily, a landowner with property flooded by a milldam could bring an action at law under the mill act. But the Essex Company's charter sought to supersede this remedy. Granting the company the same right of eminent domain given to railroads, the act required affected property holders to apply to county officials within three years for redress. Ruling for the company, the court dismissed the case because the plaintiff failed to seek relief through the proper channels.[43]

In rendering its opinion, the court equated public use with economic development. The company's dam contributed to the larger goal of economic growth – a point that justified the sacrifice of certain property rights, both for land and for fishing. But not everyone agreed with this view. Although it may have been fair to allow property owners affected by railroads three years to appeal for damages, some people believed otherwise in the case of dams. Storrow himself questioned whether the company was liable for damages for three years from completion of the dam, as he hoped, or further into the future. As he pointed out to his lawyer, Benjamin Curtis, "the flowage in freshets will be greater in some years than in others. This is independent of our control." It could of course happen, he went on, "that four years after the completion of the dam, an

---

[41] Morton J. Horwitz, *The Transformation of American Law, 1780–1860* (Cambridge, Mass., 1977), 63.

[42] *Hazen v. Essex Co.*, 12 Cush. at 477–8.

[43] Ibid., 478–80.

injury, or an amount of flowage may be occasioned, or be manifest which had not previously been expected or realized."[44] Despite such worries, Curtis assured him that the only valid claims were those made within three years of the raising of the dam to its full height.[45]

There nevertheless emerged by the 1850s a concerted challenge to this rule. Royal Call, a landowner upstream in Lowell, claimed the Lawrence Dam overflowed his property and filed a complaint with county officials in 1853. Since Call brought his case more than three years after the dam's completion, the officials refused to grant any damages. The following year, the case of *Call v. County Commissioners of Middlesex* appeared before the SJC.[46] Once again, Shaw conveyed the court's decision. The authority to raise a milldam, he explained, could derive from either the mill acts, or from a specific enactment such as the Essex Company's act of incorporation. "In either case, the damage to the landowner is intended as an indemnity, not for casual or occasional damages, which may be afterwards suffered, by a freshet or flood, but for all the damage he may suffer by all the flowing which may be caused by the erection of such dam."[47] The mill acts and the rule of eminent domain were comparable in that both sacrificed old property for new property in the name of economic growth. In addition, they both cut off traditional common-law remedies. In this case, the court ruled that the right to apply for damages began when the Essex Company completed its dam and ran for three years.

It was a rule that Call for one refused to tolerate. In 1855, after his defeat before the court, Call had his lawyer write Samuel Lawrence. The message was blunt: "The Supreme Court having decided that claims against the Essex Company for damages caused by their dam are limited under their charter to three years from the erection of the dam, Dr. Call is about to petition the legislature for a repeal of that provision of their charter."[48] That same year, landowners on the Merrimack petitioned the legislature to repeal the three-year rule.[49] With the

[44] C. Storrow to B. R. Curtis, 27 June 1850, "Record of Legal Opinions," item 7, EC Papers.

[45] B. R. Curtis to C. Storrow, 1 Aug. 1850, item 7, EC Papers.

[46] *Call v. County Commissioners of Middlesex*, 2 Gray 232 (Mass., 1854).

[47] Ibid., 236.

[48] A. P. Bonney to S. Lawrence, 26 Feb. 1855, item 89, EC Papers.

[49] Petition to the House of Representatives by Thomas Nesmith et al., 24 Mar. 1855, and Petition to the Senate and House of Representatives by A. J. Richmond et al., 29 Jan. 1856, item 89, EC Papers.

proposed legislation before the Senate in May 1856, Charles Storrow wrote Senator George W. Devereux. A brilliant engineer, Storrow was also a superb rhetorician:

> I ask, is it right for the Legislature thus to change materially the relative rights & obligations of parties equally entitled to protection. Is it equitable to take from the Company with out their fault and against their consent, rights granted by their charter, accepted with their corresponding obligations, and sustained by the judgement of the highest legal tribunal of the State – rights not unusual in their character, but which for nearly twenty years past have been enjoyed by every Railroad Corporation in the State, by virtue of general laws, and without a complaint – and all this for the purpose of subserving no public interest and redressing no wrong.[50]

If the Essex Company's charter were changed, Storrow reasoned, so too could the charters of all railroad corporations in the state, thereby undermining the state's economy. A bill repealing the three-year limitation on damages, he warned, left other corporations no choice "but to tremble in their turn."[51]

Later that month, Devereux and his colleagues tabled the bill.[52] But the tumult over the Lawrence Dam did not die down. The dam, in terms of its sheer size alone, was unlike any the valley had seen before. And its ecological impact, especially the way it blocked the passage of fish and flooded land, made it a constant focus of opposition. On 6 June 1856, scarcely two weeks after the efforts to repeal a part of the charter had been postponed, the Massachusetts legislature struck hard at the Essex Company. It passed a bill requiring the company as of February 1857 to construct "a suitable and sufficient fishway for the usual and unobstructed passage of fish during the months of April, May, June, September and October" or be fined for failing to do so.[53] Earlier, a legislative committee had examined the Essex Company's fishway and found, according to a newspaper account, "that there is not the slightest evidence that ever a single fish had passed from the water below the dam to the water above."[54] This was of course no surprise. The original fishway had never been effective. And as far as the company was concerned, the payment of damages for fishing rights excused them from any further responsibility. In effect, the

---

[50] C. Storrow to G. W. Devereux, 8 May 1856, item 89, EC Papers.

[51] Ibid.

[52] "Massachusetts Legislature," *Boston Daily Advertiser*, 28 May 1856.

[53] Act of 6 June 1856, ch. 289, [1856] Mass. Acts & Resolves 221.

[54] "The Legislature and the Fish-way of the Essex Company," *Lawrence Courier*, 20 May 1856.

power of eminent domain operated here as well, justifying the appropriation of fishing rights. Yet the passage of this act in 1856 signified an attempt to hold the company responsible for the ecological impact of its actions.

The Essex Company asked its lawyers – Charles G. Loring and E. Merwin – to consider the constitutionality of the act. "It is painful & humiliating," they began, "to contemplate a legislative act of such obvious injustice & oppression, seemingly far better suited to a semi-barbarous state of society." In their view, the Essex Company and the state had entered in effect into a contract. The company built a fishway for the dam that conformed to the expectations of county officials; when it appeared that the fishway had failed, it compensated the owners of fishing rights as stipulated by the 1848 addition to the company charter. The company had clearly fulfilled its obligations to the state. But the passage of the new act was "in effect neither more nor less than one to pervert a dam & fishway erected by its [the legislature's] permission for the public good, & in the precise manner prescribed by it . . . into a nuisance, & to constitute its continuance a crime punishable by indictment & ruinous penalties." There seemed little question in their minds that the 1856 act was unconstitutional and void.[55]

In 1857, the company appealed to the Massachusetts legislature to repeal the bill. The Essex Company wrote that it had "contributed to the industry of Massachusetts a costly, perfect and durable instrumentality for the creation of waterpower." It had called forth a city where but the scattered homesteads of 150 persons had existed before, had provided for the passage of fish as the government required, and had otherwise behaved both reasonably and responsibly. For the legislature to now pass an act requiring a new fishway when damages had already been paid, the company implied, was tantamount to theft. The bill passed in 1856, "is an Act not to alter a charter, *but to take private property for public uses without compensation.*"[56]

Rufus Choate, the famous Whig lawyer whom the Essex Company recruited to its cause, was even more emphatic. To his mind, the legislature in 1848 had authorized the corporation to purchase the right to damage fishing interests above the dam. But to permit the company to destroy fishing rights "and

[55] Legal Opinion of Charles G. Loring and E. Merwin, 30 June 1856, "Record of Legal Opinions," item 7, EC Papers.
[56] Petition of the Essex Company to the Mass. Legislature, 5, 8–9.

then to say, after they have acquired and paid for that right, that the legislature did not authorize them to do that injury to the public interests . . . is to speak absurdly." The right of destroying fisheries acquired by the company in 1848 was in essence a form of property. "If now it is said," Choate concluded, "that the public interest in the fisheries requires this destruction, it is then clearly, the ordinary case of taking private property for public uses, and it is to be paid for."[57]

In the past, the Essex Company, through its association with the public good, had been the beneficiary of the state's power of eminent domain. Now the company found itself the target of appropriation. The state had granted, and now it threatened to take away, redefining what constituted the public good and forcing the company to conform to its will. As the Essex Company pressed its case to repeal the act, a protest movement emerged in the Merrimack valley. In early 1857, petitions opposing the company's attempt to repeal the 1856 act arrived at the statehouse in Boston. The petitioners hailed from valley towns in Massachusetts – Lowell, Haverhill, Salisbury, Andover, Amesbury, Tyngsborough – and New Hampshire, reaching as far north as Manchester.[58] Evidently, the opposition had some effect. The company's efforts proved unsuccessful, leaving the state free to bring suit against it for failing to comply with the enactment.

The case of *Commonwealth* v. *Essex Company* came before the Massachusetts SJC in 1859.[59] Among other issues, the court had to decide the constitutionality of the 1856 bill. That act was part of a long line of statutes that protected and regulated the public's right to fish in Massachusetts waters.

The Massachusetts statutes were modeled on British laws regarding fishing in rivers. Since the thirteenth century, the British Parliament had regulated local fisheries, forbidding nets of certain sizes and the obstruction of rivers where fish spawned.[60] Apart from the protection granted by the statutes,

---

[57] Petition of the Essex Company, and Other Documents, 1857, p. 22, MATH.
[58] Journal of the Mass. Senate, 11 Feb.–21 Mar. 1857, pp. 161, 187, 190, 218, 225, 234, 247, 250, 260, 278, 289, 308, 346, 353, Mass. State House Library, Archives and Manuscripts, Boston, Mass.
[59] *Commonwealth* v. *Essex Co.*, 13 Gray 239 (Mass., 1859).
[60] Gary Kulik, "Dams, Fish, and Farmers: Defense of Public Rights in Eighteenth-Century Rhode Island," in *The Countryside in the Age of Capitalist Transformation: Essays in the Social History of Rural America*, ed. Steven Hahn and Jonathan Prude (Chapel Hill, N.C., 1985), 28–9.

the British common law also considered the obstruction of fish a nuisance. A landowner whose livelihood was jeopardized by an obstruction preventing fish from passing could legally resort to self-help to fix the problem.[61] Even as late as 1806, a British court affirmed the public's right to have migrating fish ascend streams, rendering total obstruction a public nuisance.[62]

When the colonists journeyed to America, they packed the common law in among their cultural baggage. But by the eighteenth century, New Englanders – no longer faced with the abundance of fish they found on arrival – began to take stronger measures. In Massachusetts, an important series of fish acts was passed in the 1740s.[63] The first statute required anyone building a dam across a river to make a passage for fish to pass upstream and to keep it open every year during April and May.[64] Two years later, the legislature specified committees to rule on the placement and dimensions of these fishways.[65] Amended again in 1745, the law exempted mill owners from building a fishway if fish no longer spawned in a river. Otherwise, and this is the significant point, a committee could rule on whether the presence of the mill or the fish made a greater contribution to the "general benefit" of the community.[66]

At the time the SJC heard the Essex case, the law in Massachusetts had evolved somewhat from its eighteenth-century origins. The significant change dated from the early nineteenth century, when the SJC no longer considered the obstruction of streams an offense under the common law. Instead, blocking a river now became indictable under the various fish acts passed over the years.

The case of *Commonwealth* v. *Knowlton* (1807) illustrates the shift in the law.[67] Knowlton built a milldam that stopped fish from passing up the Sandy River. The state brought suit under a statute passed in 1798. That law protected salmon, shad, and

---

61 Humphrey W. Woolrych, *A Treatise of the Law of Waters*, 2d ed. (London, 1851), 192–5. A list of British fishery statutes since the thirteenth century can be found in Stuart A. Moore and Hubert Stuart Moore, *The History and Law of Fisheries* (London, 1903), ix–xxviii; also see James Patterson, *A Treatise on the Fishery Laws of the United Kingdom* (London, 1863), 96–8, 100, 103.
62 *Weld* v. *Hornby,* 7 East 195, 103 Eng. Rep. 75 (K.B. 1806).
63 For an earlier enactment, see Province Laws 1709–10, ch. 7.
64 Province Laws 1741–42, ch. 16.
65 Province Laws 1743–44, ch. 26.
66 Province Laws 1745–46, ch. 20.
67 *Commonwealth* v. *Knowlton,* 2 Mass. 530 (1807).

alewives by forbidding the obstruction of rivers in Lincoln and Cumberland counties. In his defense, the defendant claimed the state had not properly charged him with creating a nuisance – a common-law offense – as the statute required. Was blocking a river an offense under the common law or under the recently passed fish act? Interpreting the obstruction of fish as a common-law offense, the court pointed out, would result in "much inconvenience." As they explained: "Every dam not having a sufficient passage-way for fish being a common nuisance, any individuals, of their private authority, might, before indictment or conviction, undertake to remove it at any season of the year. Resistance would be the probable consequence, the certain effect o[f] which would be a breach of the peace, attended with outrage and violence."[68] Construing the obstruction of a river as an offense against the fish act kept the peace by preventing the affected party from resorting to self-help to abate the nuisance. In the court's opinion then, the defendant's actions were more properly indictable under the newly created fish statute.

A more explicit statement of this view emerged in *Commonwealth v. Chapin* (1827), another case in which the defendant blocked the passage of fish with his dam.[69] In giving the court's opinion, Chief Justice Parker affirmed the upstream riparian landowner's right against his downstream neighbor to have fish ascend the river. Invoking the cautionary language of the common law – *sic utere tuo, ut alienum non laedas* (use your property so as not to harm that of another) – Parker seemed convinced that totally obstructing the passage of fish constituted a public nuisance. Yet he also noted how the fish acts passed by the state had effectively superseded the common law. The public's right to have fish pass up its rivers, he announced, is "to be determined by the effect and according to the form of this legislation, rather than by the ancient common law." Merely constructing a dam that blocked fish was no longer deemed an offense. "The offence," as Parker put it, "consists only in having a dam without providing a convenient passage for the fish during two or three months in the year." Moreover, since a failure to comply was no longer a nuisance under the common law, the traditional avenue of relief through abatement was cut off. The fish

---

[68] Ibid., 534.
[69] *Commonwealth v. Chapin*, 5 Pick. 199 (Mass., 1827).

acts instead applied penalties and appointed officials "to supervise the public interests, and see to the execution of the law."[70]

By the second quarter of the nineteenth century, regulation of inland fisheries no longer fell within the scope of the common law.[71] Instead, legislative statutes protected the public's right to fish. The change left affected parties to face a rather different set of alternatives for relief. But it also substituted statutes, which legislated for the common good on an ad hoc basis, for the universal authority of the common law. In short, the legislature was to decide how to regulate inland waterways in the public interest.[72] Like the mill acts discussed earlier, this trend in the law opened the way for economic development. It did so by letting legislators weigh the relative merits of obstructing rivers for waterpower purposes against the preservation of fish. The state government could now legislate in favor of economic growth, but it decided the fate of the region's ecology in the process.

When Chief Justice Shaw composed the opinion in *Commonwealth v. Essex Company,* it was law, not ecology, that concerned him. Narrowly speaking, the court had to decide whether the 1856 act violated the contract entered into by the company and the state in 1848. But the court also spoke to the broader issue of state regulation of inland fisheries. He began by considering the present state of the law. "It seems to be well settled," Shaw announced, "that the obstruction of the passage of the annual migratory fish through the rivers and streams of the commonwealth, is not an indictable offence at common law."[73] Echoing the prevailing legal precedent, he noted that the public's right to the passage of fish was regulated by statute.

Shaw next discussed the significance of the Essex Company's 1845 charter. "The objects proposed to be accomplished by the defendants were so far public in their nature," he intoned, ". . . that it was quite competent for the legislature to exercise

---

[70] Ibid., 202, 204.

[71] See Joseph K. Angell, *A Treatise on the Law of Watercourses,* 5th ed. (Boston, 1854), 84, n. 2. James Kent explains that in Massachusetts, the obstruction of a river subjects the violator to penalty under statute. He also points out that in Connecticut, similar statutes had been passed. "But manufacturing machinery, and steamboats, and the insatiable cupidity and skill of fishermen, have prodigiously diminished the resort of the most valuable fish into the rivers of the northern states." James Kent, *Commentaries on American Law,* 6th ed., 4 vols. (New York, 1848), 3:411, n. a.

[72] On the Rhode Island fish acts, see Kulik, "Dams, Fish, and Farmers," 30–1.

[73] *Commonwealth v. Essex Co.,* 13 Gray at 247.

the power of eminent domain."[74] Had the additional act of 1848 not been passed, Shaw implied that the 1856 enactment might well have been a legitimate amendment to the company's charter.[75] Yet the 1848 act allowing the company to compensate the owners of fishing rights changed matters. That act substituted a cash payment for the damages incurred by the failure of the Lawrence Dam fishway. In Shaw's view,

> the legislature had the power to regulate the public right, and diminish it or release it, as the best good of the public, on the whole, might in their judgment require. Whether that public good, expected from the fishery, consisted in affording an additional article of food to the people, or an employment for labor, or otherwise, the legislature might well compare this with the public advantage, in affording increased profitable labor and means of subsistence, and various benefits, from building up a large manufacturing town, and decide as the balance of public benefit should preponderate.[76]

In the end, Shaw believed the 1848 act had "all the elements of a contract, executed by one party and binding on the other."[77] For this reason the subsequent enactment passed in 1856 was unconstitutional. In 1848, the state essentially exempted the company from making a fishway by providing damage payments. Eight years later it was nullifying the exemption and leveling new obligations on the company. The government, Shaw believed, had acted on behalf of the public by excusing the Essex Company from maintaining fishways and providing indemnities instead. To alter or amend this contract by rescinding the exemption meant crossing the fine line into the unconstitutional.[78]

But Shaw's decision begs the question, Did the 1848 act extinguish *all* public interest in fishing? Were there any public rights to fish that remained after the Essex Company paid compensation to the fisheries? The 1856 act can be understood to say that the payments did not extinguish potential future rights in the public to fish. Shaw, however, held that the public right consisted only of the existing private rights to fish compensated for in 1848. In this sense then, Shaw narrowed the public's interest in fishing rights to the benefit of waterpowered industry.

---

[74] Ibid., 249.
[75] See Leonard W. Levy, *The Law of the Commonwealth and Chief Justice Shaw* (1957; reprint, New York, 1967), 278.
[76] *Commonwealth v. Essex Co.*, 13 Gray at 251.
[77] Ibid., 252.
[78] Ibid., 253–4.

The Essex case also forced the Merrimack valley further down a particular ecological path. That path was pioneered by the Boston Associates and others who threw dams in the way of fish swimming upstream. Well before the completion of the Lawrence Dam, the Merrimack fisheries were far along the path of ecological ruination – if not caused by obstruction then by overfishing. The Lawrence Dam finished off this new chapter in the valley's ecological history, completing the devastation of the watershed's fisheries. With the construction of the dam, the notion of public good had been further redefined to favor the industrial use of the river's water. State regulation of the river had resulted in a rather different ecological calculus, one that, if the 1856 legislation is any indication, some found disquieting. Capital, technology, and law combined in the nineteenth century to bring industrial development to the valley. But economic growth spelled ecological transformation as well. The task ahead was to recover that old ecology, to restore the valley back to the days when spring sent Matthew Patten and his neighbors, flush with anticipation, to the banks of the Merrimack.

New Englanders hitched their hopes for restoring fish to the region on artificial propagation. Fish culture, as artificial propagation was often called, involved securing fish eggs of depleted species, hatching them under controlled conditions, and reintroducing the young fish into streams or lakes. It appealed to New Englanders, who viewed it as a reasonably simple way of resurrecting the region's inland fisheries. More generally, it signified a deeper, more basic faith that science and technology could counteract the less desirable ecological costs of industrial change. It tried to recreate a past ecology, to redesign the natural world, yet again, to suit the pressing concerns of the moment. In the Merrimack valley, where the environmental impact of economic development had grown dreadfully apparent, fish culture attracted support from politicians, scientists, commercial fishermen, and farmers who often saw it as another opportunity to make the region's waters commercially successful.

Although the techniques of artificial propagation had been discovered earlier, its birth as a commercial endeavor came in the nineteenth century. France emerged as the center of scientific innovation. During the late 1840s and into the next decade, the methods for mass cultivation of fish eggs were developed and eventually perfected. Such advances combined with the evolution of the science of ichthyology – the branch of zoology concerned with fish – made artificial propagation a practical reality. By midcentury, fish culture was a budding commercial enterprise. One French professor active in the pioneering efforts wrote in 1850, "There is no branch of industry or husbandry, which, with less chance of loss, offers an easier certainty of profit."[79]

The progress in France eventually filtered across the Atlantic. In 1856, Massachusetts appointed a commission to investigate the artificial propagation of fish.[80] One member of the commission tried to cultivate trout eggs on Cape Cod but met with only limited success. Yet the commissioners remained hopeful. As they put it, artificial propagation was "not only practicable but may be made very profitable."[81] They saw fish cultivation as a viable commercial venture with promise for ordinary people. They envisioned individuals with modest amounts of capital and skill, with help from the fish culture manuals then available, restocking the state's smaller streams and ponds.[82] Larger rivers, such as the Merrimack and Connecticut, they agreed, would require the combined efforts of a number of people, including special legislation to guard "the rights of all persons interested in the waters, especially when they have been applied to mechanical purposes."[83]

The ultimate success of fish culture in the Merrimack valley

[79] Quoted in Mass. Legislature, Committee on Artificial Propagation of Fish, "Report" (May 1857), S. Doc. 193, 41. The best short review of the history of fish culture is Jules Haime, "Pisciculture," *Revue des Deux Mondes*, 15 June 1854. It appears in translation in public documents that consider the viability of fish culture in New England. See the 1857 report noted, 19–54; also see M. Coste, *Instructions Pratiques sur la Pisciculture Suivies de Mémoires et de Rapports sur le Même Sujet* (Paris, 1853), 1–21.

[80] Resolve of 16 May 1856, ch. 58, [1856–7] Mass. Acts & Resolves 277. Other New England states, including Connecticut and Vermont, also explored fish culture. See Resolve of 17 Nov. 1856, No. 101, [1853–6] Vt. Acts 112. George P. Marsh, *Report, Made Under Authority of the Legislature of Vermont, on the Artificial Propagation of Fish* (Burlington, Vt., 1857), 41–8, discusses fish culture efforts in Connecticut during the 1850s.

[81] Mass. Legislature, Committee on Artificial Propagation of Fish, "Report," 12–13. See pp. 14–17 where N. E. Atwood, the commissioner who experimented with breeding trout, explains his endeavors.

[82] An example of one of the earlier fish culture manuals is W. H. Fry, ed. and trans., *A Complete Treatise on Artificial Fish-Breeding* (New York, 1854). This book contains a translation of M. Coste, *Instructions Pratiques sur la Pisciculture*.

[83] Mass. Legislature, Committee on Artificial Propagation of Fish, "Report," 13.

depended on an unobstructed river. Before establishing hatcheries to produce fish eggs, fish needed to have access to spawning grounds upstream where they could eventually reproduce on their own. The successful return of migrating species of fish hinged on their free passage upstream. Yet the problem sat unaddressed for a number of years. The initial impetus for making the Merrimack River accessible to migrating fish eventually came from New Hampshire. Noting the decline of salmon, shad, and other species in the Connecticut, Merrimack, and Saco rivers, a New Hampshire bill, passed in 1864, requested Massachusetts, Connecticut, and Maine to take steps to restore free passage up the region's rivers.[84] The following year, New Hampshire's governor, Joseph Gilmore, criticized the efforts of Massachusetts to deal with the matter:

> The importance of fishways to give a free passage to fish through artificial dams to the waters of this State, and the obligation of Massachusetts to construct them has been acknowledged and recognized by that State in her acts incorporating the Essex and Lowell manufacturing companies. The dams of these companies have been so constructed, however, as to render the passage of fish an impossibility.[85]

The intrusion of the mills along the lower Merrimack into New Hampshire had been a source of trouble since the 1850s. Now such anger mingled with concern over the environmental impact of the Lawrence Dam. That point is evident in an 1866 conversation between the Lake Company's Josiah French and Frederick Smyth, New Hampshire's newly elected governor. Smyth lectured French on the failure of Massachusetts to address seriously the need for fishways so that fish could once again ascend into the waters of his state. The governor threatened that while New Hampshire could not compel fishways to be built in Massachusetts, it could pass legislation affecting the dams that barred the fish. As French recalled, the governor warned him of the

> strong feeling in some parts of the state in relation to the management and control of so much of the navigable water of the state and that he thought some legislation would be demanded to protect New Hampshire interest which seemed to be entirely subservient to that of Massachusetts.[86]

---

[84] Resolve of 16 July 1864, ch. 2898, [1862–6] N. H. Laws 2854.
[85] Reprint of Governor Joseph Gilmore's Address to Legislature, *New Hampshire Patriot and State Gazette*, 14 June 1865.
[86] French recounts his discussion with Smyth in J. French to J. Francis, 12 Apr. 1866, vol. A-39, file 216, PLC Papers.

French countered by explaining how the Lake Company had improved the waterpower of both New England states. He pointed out "that although Massachusetts received some distinct benefit and did control the water it was not done to the detriment of New Hampshire's interest in the use of the waters, but largely to its benefit."[87] That argument had been used before but was wearing thin. The waters of New England were unquestionably more valuable to industry because of the efforts of the Waltham–Lowell mills. But just as certainly, the industrial control of water resulted in ecological costs, a lesson not lost on New Hampshire's political leadership. Finding the state's waters depleted of fish, they pushed for legislation across the border to address the problem.

In response, Massachusetts appointed a congressional committee to consider building fishways along the Merrimack and Connecticut rivers. Judge Henry A. Bellows (the same man who ruled in *Lake Company v. Young*) addressed the committee in 1865 on New Hampshire's behalf. Admitting the importance of manufacturing to the Merrimack River, Bellows implored the committee not to let this fact obscure his state's right to catch fish as they passed upstream.[88] The committee shared his concerns, noting that "the appropriation of the rivers and ponds to manufacturing purposes, the magnitude of the interest, and the success . . . has withdrawn the public mind from that solicitude for the preservation of fish which formerly engaged so much attention here."[89]

To help them understand why the region's fisheries had declined, the committee consulted the eminent naturalist Louis Agassiz. He blamed the devastation of fisheries not on dams, but on factories that used rivers to dispose of industrial wastes such as sawdust, dyes, and acids. Drawing the attention of the committee to France, where artificial propagation had restored fish stocks without compromising manufacturing, Agassiz advised New Englanders to follow a similar course. In his optimistic view, Europe's experience suggested that fishways could be effective and fish culture successful.[90]

When James B. Francis came before the committee, he had of course a somewhat different perspective. The Lowell mills, he

---

[87] Ibid.

[88] Mass. Legislature, Joint Special Committee on the Obstruction of the Connecticut, Merrimack, and Saco Rivers, "Report," 4.

[89] Ibid., 10–11.

[90] Ibid., 5–7.

pointed out, required all the water the Merrimack could supply, especially in the spring and summer. To leave open a fishway between April and June to insure the free passage of fish would come at the expense of manufacturing. Despite his hesitations, however, he conceded that with enough money a suitable fishway might be built.[91] Francis's words weighed heavily on the committee. As they wrote, "The suggestion that the mills at Lowell use substantially all the water at Patucket [*sic*] Falls, and that a fish-way could not be kept open without affecting the supply of water under the present arrangement at a later period in the year . . . calls for a thorough examination into the expediency of any legislative action in the premises."[92]

Whatever might be done at Lowell, the far more pressing question concerned the formidable Lawrence Dam. The committee freely admitted that the failure to construct a proper fishway at Lawrence – the error of the county officials who mistakenly approved the flawed construction – clearly injured the rights of New Hampshire. Fortunately, Charles Storrow, who testified as well, had no objection to a new fishway if it did not jeopardize the security of the dam.[93] The committee agreed that the error at Lawrence should be rectified so long as "it can be done without sacrificing the greater interests which are depending upon the use of the water for manufacturing purposes."[94]

Later that year a far more thorough examination of inland fisheries took place, this one directed by Theodore Lyman.[95] An aspiring zoologist who had worked with Agassiz, Lyman was to play a pivotal role in the attempt to restore fish to Massachusetts waters.[96] Lyman, together with his colleague Alfred Reed, were appointed to examine the obstructions to fish in the Connecticut and Merrimack rivers. On the Merrimack, they found the supply of water at the Lawrence Dam abundant enough to construct a fishway without compromising manufacturing. At Lowell, however, the dam was too low to provide water for all the mills, and a fishway would aggravate the problem. A water shortage at Lowell, their report concluded, had serious implications: "It not

[91] Ibid., 17.
[92] Ibid., 21.
[93] Ibid., 18.
[94] Ibid., 22.
[95] Mass. Report on Obstructions.
[96] Allen Johnson and Dumas Malone, eds., *Dictionary of American Biography*, 22 vols. (New York, 1936–58), 6:519.

only touches the capitalist, but immediately affects between eleven and twelve thousand operatives. If, from slack water, the machinery runs slow, those operatives who work by the piece earn little, and are discontented."[97] Still, they believed enough water flowed there to operate a fishway in the spring. It was agreed that restoration of migrating fish must not cut too deeply into the interests of the valley's great waterpowered manufacturing. No one proposed to level any dam to bring back the region's former ecological plenitude.

Lyman and Reed also considered how pollution threatened the river's fish, a point made by Agassiz.[98] As industry flourished in the nineteenth century, the Merrimack River served as a convenient site for disposing of industrial wastes – a significant source of water pollution in the valley by the 1860s. (The valley's water quality is considered more fully in Chapter 7.) Although their report noted the "notoriously pernicious" effects of sawdust on fish, they showed more concern for the pollutants released by the printing and dying done at Lawrence.[99] The pollution below the Lawrence Dam became severe, the report noted, after the Pacific Mills began operating in 1856, "a factory that uses, perhaps, fourfold more dyes and chemicals than do all the other factories in the place together" (Figure 6.2). As a result, the shad fishery below the dam was devastated the following year. The Essex Company had customarily leased fishing rights below its dam, taking in an average revenue of close to three hundred dollars per year between 1850 and 1857. But during 1857, after the Pacific Mills started operating, fishermen found the quantity of shad below the dam vastly diminished. When pollution damaged the shad fishery in 1857, fishermen refused to rent fishing rights from the company, and the annual rental revenue fell to just seventeen dollars in the following year. In the period from 1858 to 1865, revenues dropped sharply to an average of twenty-six dollars per year.[100]

Streams and rivers were considered at this time legitimate depositories for wastes, and the report made no attempt to bar

[97] Mass. Report on Obstructions, 12.

[98] The interest of fish culturists in pollution is explored in Donald J. Pisani, "Fish Culture and the Dawn of Concern over Water Pollution in the United States," *Environmental Review* 8 (1984): 117–31.

[99] Mass. Report on Obstructions, 16, 18–20.

[100] Ibid., 19–20.

Figure 6.2. The mammoth Pacific Mills and Print Works. (Courtesy of the Museum of American Textile History.)

factories from polluting water.[101] But not all pollution was equally excusable. "In a word," the report explained, "a fair stream is a mechanical power and a lavatory, but it is *not* a common sewer."[102] That thought cut to the core of the nineteenth-century view of water as integral to the production process. As a source of waterpower and a convenient place to dispose of factory waste, rivers were indispensable links in production. The word lavatory is used here in its broadest sense, as a vessel used for washing. But the indiscriminate practice of tossing pollutants unconnected with manufacturing (excluding

[101] Erecting wooden barriers in the river opposite the raceways of mills, the report suggested, would keep the polluted water separate from the main body of the stream. "Prevented, in this way, from flowing into the middle of the river, it will, after a run of a few hundred feet, probably, become pure enough to mingle with the rest of the water." See ibid., 25. The following year, the Pacific Mills agreed to place such plank fences at each of its raceways. See, Mass. Commissioners of Fisheries, *First Annual Report* (1867), H. Doc. 3, 5. Such action was based on the belief that running water purified itself. Readers can find a full discussion of this issue in Chapter 7.
[102] Mass. Report on Obstructions, 21.

unavoidable human wastes) into a river – using it as a "common sewer" – could not be condoned. The line dividing right from wrong was both calculating and obsessive, bent on preserving water for the higher purposes of production.

Preventing unnecessary pollution and constructing fishways, the report concluded, were key steps in restocking the Connecticut and Merrimack rivers with shad and salmon. They recommended further that New Hampshire be encouraged to breed salmon artificially and that laws be passed to prevent overfishing. "If the above conditions were complied with," Lyman and Reed wrote somewhat cautiously, "an abundant supply of fish might reasonably be looked for within five years, though they would not be so plenty as when the country was in its primitive state."[103]

The report culminated with the passage of a major piece of legislation the following year. That legislation appointed an official Massachusetts commission to oversee the restoration of fish to the Merrimack and Connecticut rivers. It empowered the commissioners to determine the need for fishways, to supervise their construction, and to dictate when they were to remain open. The act also furnished seven thousand dollars in state funds to build a new fishway at the Lawrence Dam (the Essex Company was to pay any additional expense). To help with enforcement, stiff fines were to be levied for noncompliance.[104] The new law won no sympathy from the waterpower companies at Lowell and Lawrence. As James Francis confided to Charles Storrow: "Under ordinary circumstances I should deem it very objectionable to us, but I think it is as well as we can expect."[105] The pressure for action was indeed extraordinary, with much of it emanating from New Hampshire. There the legislation was hailed in triumph by Governor Frederick Smyth.[106]

A systematic campaign to restock New England's waters with fish had emerged by 1867. At that date, five states – Maine, Massachusetts, Connecticut, Vermont, and New Hampshire – had all formed fish commissions to supervise the restoration of migrating fish to the region's inland waterways. In the Mer-

---

[103] Ibid., 40–1.
[104] Act of 15 May 1866, ch. 238, [1866] Mass. Acts & Resolves 231.
[105] J. Francis to C. Storrow, 21 Apr. 1866, item 14, EC Papers.
[106] *Message of His Excellency, Frederick Smyth, Governor of the State of New Hampshire to the Two Branches of the Legislature* (Concord, N.H., 1866), 39.

rimack valley, Massachusetts enacted legislation making fishways mandatory. Although New Hampshire law had tried diligently to promote unobstructed rivers, those laws were repealed by 1842.[107] But progress was made in 1865 when the
state passed an act requiring fishways along the state's major
rivers including the Winnipesaukee, Pemigewasset, and Merrimack.[108] In 1867, both states also prohibited taking salmon
and shad (and alewives in Massachusetts) from the Merrimack
for at least four years (five years in Massachusetts).[109]

By prohibiting fishing, the states hoped to give their restocking efforts a chance to thrive. In 1866, Atlantic salmon eggs
obtained from the Miramichi River in New Brunswick, Canada,
were planted in the Pemigewasset River. With the backing of
New Hampshire and Massachusetts, Livingston Stone – a key
figure in the fish culture movement – opened a hatchery for
salmon eggs on the Miramichi in 1868. Beginning in 1871, a
more permanent supply of salmon eggs was shipped to the
Merrimack from Maine's Penobscot River, where salmon could
still be found naturally.[110] While salmon were being re-
introduced upstream, Massachusetts undertook the artificial
propagation of shad in North Andover. In 1869, over 2 million
shad eggs were sent from there to the region's waters – some as
far as Lake Winnipesaukee – while more than 2.6 million ova
were turned into the Merrimack alone.[111]

Fish culture was often touted as a potentially lucrative financial enterprise. Theodore Lyman and Alfred Field pointed to
uninformed New Englanders who persisted in "putting up
more thousands of spindles, and flooding the market with un-
salable cotton goods, when, from the very water which turns
their wheels, they might coin money, with no other machinery
than a net and a hatching trough!"[112] As they explained, fish
culture was really little different from the cultivation of grain.
To the hesitant, those wary of tampering with nature, they

[107] New Hampshire laws regarding the obstruction of fish in the Merrimack valley are
discussed in the records of a North Andover textile manufacturer. See Daniel
Barnard, Memo on Fishways, item 18, Stevens Companies Papers, MATH.

[108] Act of 1 July 1865, ch. 4099, [1862–6] N. H. Laws 3136.

[109] Act of 14 June 1867, ch. 1, [1867–71] N.H. Laws 1; Act of 31 May 1867, ch. 289,
[1867] Mass. Acts & Resolves 688.

[110] Stolte, *Forgotten Salmon of the Merrimack*, 16–17, 21. Stone later became secretary of
the American Fish Culturists' Association. See McEvoy, *Fisherman's Problem*, 105.

[111] Mass. Commission on Inland Fisheries, *Fourth Annual Report* (1870), S. Doc. 12, 7–
8.

[112] Mass. Commission on Fisheries, *Second Annual Report* (1868), H. Doc. 60, 19.

asked rhetorically: "What advantage is there in sowing, over wild growth? Why not let corn grow and sow itself in its own way?"[113] If New Englanders had for centuries plowed and harvested the land, there seemed no reason why – with a little bit of diligence and ingenuity – water could not be forced to yield its own kind of crop. Fish culture was offered as a new frontier, an innovative way of prospering from the New England waterscape.

Moreover, the emergence of artificial propagation demanded a rather different understanding of fish as a form of property. Massachusetts sought to restructure the law to conform more nearly to the new economic potential of fish culture. For over two hundred years, a colonial ordinance (1641–7) gave every "householder" in the province the right of free fishing in bays, coves, and rivers (at least in rivers deemed navigable and therefore public), unless otherwise appropriated by a town. Since no town could convey a pond of more than ten acres into private hands, the law made all ponds over this size the common property of all.[114] With no pressing reason to repeal the law, it remained as one of the last vestiges of a legal code adopted in a time when production was less of a concern. Yet the advent of artificial propagation sparked a call for reform.

The time had come, according to Lyman and Field, to extend the laws of private property to include fish. In the past, they reasoned, other aspects of the natural environment such as land and wood had become subject to private property.[115] Indeed, they argued in 1869 that private property gave a vast incentive to the forces of economic development:

> If sheep, cattle, and horses could be legally killed, or caught and sold, by the first passer-by, whenever they were kept in a field exceeding ten acres, or whenever they chanced to stray into a neighboring road or pasture, it is not likely that these useful animals would be much bred. . . . In point of fact we are trying to live under laws that applied, not to a rich, densely populated and intensely commercial people, but to a feeble community of scattered habitations, without great industries, and relying partly on game to eke out the scanty products of the newly-cleared forest land.[116]

113 Ibid., 13.
114 Herbert E. Locke, "Right of Access to Great Ponds by the Colonial Ordinance," *Maine Law Review* 12 (1919): 148–9; James Sullivan, *The History of Land Titles in Massachusetts* (Boston, 1801), 284.
115 Mass. Commission on Fisheries, *Second Annual Report*, 7.
116 Mass. Commission on Fisheries, *Third Annual Report* (1869), S. Doc. 3, 21–2.

Fish culture had now become a new industry, "one that should bring wealth to individuals and revenue to the Commonwealth." By changing the law and allowing individuals the exclusive right to fish in certain areas, the state's inland fisheries could "*be thrown open* to the energy of private persons." They imagined ultimately that "each pond and brook might team with the best trout, black bass, white fish, land-locked salmon, etc., if only they were recognized as property."[117]

In June 1869, the legislature approved a bill that encompassed this new understanding of property. Titled "An Act for encouraging the Cultivation of Useful Fishes," the law codified the rules governing the state's fisheries. A massive piece of legislation (with thirty-four sections in all), the act, among other things, gave riparian proprietors the right to exclusive control of any pond of less than twenty acres, disposing of the colonial statute and doubling the allowable size. Although ponds of more than twenty acres were to remain in the public domain, the new law gave the fish commissioners the right to lease such waters (twenty-six of the state's great ponds were leased by 1874) for fish culture. The lessees had an exclusive right to all fish produced artificially in these ponds.[118]

The new law suited the needs of the still fledgling fish culture industry. Following its passage, the Massachusetts fish commissioners commented on the significance of the act. The state's supply of freshwater fish was disappearing, they noted, and in some locales certain species were already extinct. "For this deplorable state of things," they insisted, "there is but one remedy, and that is to make fishes, under certain conditions, PROPERTY, and thus give the same stimulus to the cultivation of fish, that is given to the raising of any other live stock."[119] That rationale was the cornerstone of the new law. The law aimed to encourage production by offering to protect the property of those who practiced fish culture. Admittedly, the act had not been designed to restore the state's former abundance of fish. But in truth, the decline of inland fisheries lamented by the commissioners had resulted at least in part from the appropriation of rivers for private gain. How fitting that the attempt to restore the region's fish stock hinged on placing water more firmly in private hands.

[117] Ibid., 22, 25.
[118] Act of 12 June 1869, ch. 384, [1869] Mass. Acts & Resolves 677.
[119] Mass. Commission on Inland Fisheries, *Fourth Annual Report,* 42.

Several years later, Theodore Lyman reflected again on the 1869 law. His thoughts had scarcely changed. Indeed, he continued to believe that by extending the scope of private property – making it easier for proprietors to maintain exclusive control over both water and fish – the state's supply of fish would bounce back. To his mind, the 1869 bill was a clear and cogent response that simplified the state's policies for inland fisheries. He admitted, however, that the bill as passed had failed to include three principal points under consideration: that fish commissioners be empowered to stop the discharge of pollution from factories into rivers and ponds, that no one pollute the waters of an already existing fishery, and that it be declared illegal for anyone to have fish in their possession during the season when they were off limits. Lyman felt that the first point was wisely excluded since "the power to stop pollution is too extensive and too complex for fishery commissioners to hold." The second point, he continued, might well have remained. But it was the defeat of the third measure that he termed a "capital error." As he noted, it was impossible for anyone to "go to a market and prove that certain fish were taken *within the Commonwealth.*" And since no one could know for sure whether a fish was taken from waters in Massachusetts or elsewhere, the opportunity for illegal sales remained.[120]

Illegal fishing, as Lyman knew, plagued the efforts to restock the Merrimack. As noted, both Massachusetts and New Hampshire enacted laws banning fishing in the river for several years. According to the Commission on Inland Fisheries, the Massachusetts act "closing the river did not prevent fishing, for many of the fishermen were not disposed to comply with any regulations."[121] Moreover, as the efforts at restoring fish were showing some signs of success, the restrictions were lifted. In 1874, Massachusetts allowed fishing for shad and alewives in the Merrimack three days a week between 1 March and 10 June.[122] The year following, the Massachusetts fish commissioners found "considerable violation of the law," as fishermen snagged, in their estimation, no less than twenty thousand shad

[120] Mass. Commission on Inland Fisheries, *Tenth Annual Report* (1876), S. Doc. 24, 49, 62–4. In 1900, the U.S. Congress allowed states to regulate wildlife brought within their borders according to their own laws and statutes. See McEvoy, *Fisherman's Problem*, 115–16.

[121] Mass. Commission on Inland Fisheries, *Tenth Annual Report*, 14.

[122] Act of 7 Apr. 1874, ch. 144, [1874] Mass. Acts & Resolves 99.

in the spring of 1874. Lyman and his colleagues on the commission believed the success of the state's fish culture program called for more than simply new laws. It required a much deeper commitment on the part of the state's citizens, a pledge of "earnest cooperation, and, to some extent, the subordination of self to the public good."[123]

Unfortunately, the state's waters continued to breed temptation. As the commissioners pointed out, fishermen below the Lawrence Dam appeared "strongly imbued with the idea that all fish allowed to pass their seines will be caught by their neighbors above."[124] Here was the perennial dilemma of the river commons. It posed a problem that threatened the still fragile success of the region's fish restoration efforts. Fishermen, the frustrated Massachusetts commissioners explained, "are the only class of men who appear to be blind to the future, and rigidly practise the doctrine of taking no thought for the morrow."[125] Despite the best efforts of the state to subject water to private control, it proved virtually impossible to compel cooperation. Placing the waters of the state more squarely in private hands, if anything, worked against the goals of fishery restoration. The Massachusetts fish commissioners had shown themselves solid supporters of private property. Faced with managing the restocking of the state's waters, they actively encouraged the use of that water for personal gain. To expect altruism and cooperation seemed out of line with this agenda.

Those who fished were of course not the only ones seeking to profit off the waterscape. Ultimately, the success of the restoration effort demanded the cooperation of the Merrimack valley's waterpower giants. However reluctant at first, the region's vast manufacturing interests complied with the fish restoration efforts. At least this was the opinion of the Massachusetts commissioners. Mill owners, they wrote in 1879, "who above all others were supposed to have the best reason for opposing it [the state's fish culture program], have in this State borne themselves in most instances with commendable good taste, and in some instances with great liberality."[126] In New Hampshire,

---

123 Mass. Commission on Inland Fisheries, *Ninth Annual Report* (1875), S. Doc. 26, 12, 21.
124 Mass. Commission on Inland Fisheries, *Tenth Annual Report*, 14.
125 Mass. Commission on Inland Fisheries, *Eleventh Annual Report* (1877), S. Doc. 8, 16.
126 Mass. Commission on Inland Fisheries, *Fourteenth Annual Report* (1879), Public Doc. 25, 16.

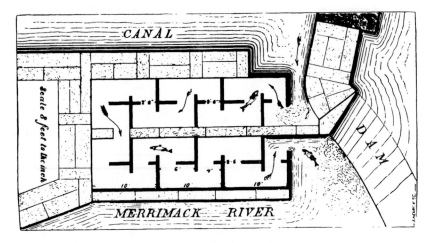

Figure 6.3. A proposed plan for a Brackett-style fishway to be built at Amoskeag Falls. (From *Report of the Fish Commissioners of New Hampshire,* 1878.)

industry proved only slightly less supportive. Although the Franklin Falls Company, which owned four dams on the Winnipesaukee River, refused at first to build fishways, the state used legal action to compel it to cooperate.[127] Further up the river, the Lake Company, reluctant no doubt to stir up any further trouble, by 1868 had erected fishways at all of its dams. On the Merrimack itself, fishways were in place by 1868 at Lawrence, Lowell, and Manchester.[128]

During the 1870s, however, the Merrimack River fishways were reconstructed. The original structures, based on a model designed by Nathan Foster of Maine, were built from the river up to the dam's crest. Separate bays inside the fishway, with wooden cross-pieces set at an angle within each compartment, sought to deaden the current. This design did not work well. Observers noticed that fish entered the first bay, but because the current proved stronger in each successive bay, many fish were unable to ascend the fishway. In the 1870s, E. A. Brackett, a Massachusetts fish commissioner, designed a new fishway (Figure 6.3). His fishway looked much like the earlier one, but extended through the dam to the water in the river above, with

---

[127] N.H. Commission on Inland Fisheries, *Annual Report* (1869), 650, and *Annual Report* (1871), 6. Also see *State v. Franklin Falls Co.*, 49 N.H. 240 (1870).
[128] N.H. Commission on Inland Fisheries, *Report* (1868), 7.

several gates located at this point. Depending on the height of the water in the mill pond above the dam, the gates were used to control the water entering the fishway. This allowed water to be let into the fishway even if the water level of the river happened to be low.[129] Brackett's fishways were also more successful than the earlier ones in breaking the river's powerful current, enabling fish to move more easily from bay to bay and over the dam.[130] By the end of the 1870s, the new fishways at all three Merrimack valley textile cities had met the approval of New Hampshire's fish commissioners.[131] Still straddled by a vast amount of technical apparatus, the river was once again open for fish to pass.

In 1878, the New Hampshire fish commission trumpeted the return of salmon to the upper reaches of the Merrimack and Pemigewasset rivers.[132] The long awaited homecoming of Atlantic salmon – a wait that stretched across decades – appeared to demonstrate the triumph of human ingenuity. Five years earlier, 174,000 salmon fry (young salmon) were released into the Pemigewasset River.[133] As is common for the species, they remained in the river's freshwater for a couple of years, journeyed downstream to the sea where they increased in size, and returned two years later to the river's headwaters. The large number of fry introduced in 1873 coincided with the successful completion of fishways at the major dams, fueling the first significant spawning run in decades. Evidently the combined efforts of New Hampshire and Massachusetts had succeeded. In 1877, New Hampshire constructed a hatchery, which made it no longer dependent on neighboring Maine for fish eggs.[134] Future prospects for restoring the river's past ecology seemed bright for the moment.

Yet darker signs were eager to intrude. Despite the significant energy poured into artificial propagation, the river's stock of shad remained tenuous. The warnings of the Massachusetts

---

129 Mass. Commission on Inland Fisheries, *Seventh Annual Report* (1873), S. Doc. 8, 18–19. Also see the plans for both the Foster and Brackett fishways in Supplement D of the report, p. 30.

130 For more on the success of the Brackett fishway installed at the Lawrence Dam, see N.H. Commission on Inland Fisheries, *Annual Report* (1877), 28.

131 N.H. Commission on Inland Fisheries, *Annual Report* (1870), 4–5; (1873), 4; (1876), 12; (1877), 7; (1878), 10–11; (1879), 3; (1880), 4.

132 N.H. Commission on Inland Fisheries, *Annual Report* (1878), 3.

133 Stolte, *Forgotten Salmon of the Merrimack*, 39–40.

134 Ibid., 61–3.

commissioners to refrain from illegal fishing had gone unheeded. In 1878, the commissioners noted that fishermen had "so persistently . . . swept the river during the last three years, that very few spawning shad have been left to keep up the stock."[135] Most of the overfishing occurred below the Lawrence Dam. Clever New Englanders could reengineer the environment and restore the fish that industrial development had chased from the waterscape. They could not, however, put an end to the dilemma posed by a river commons. State regulation of New England's waters had never been terribly effective. The sheer abundance of legislation passed throughout the state – bills enacted, disputed, repealed – is testimony enough. Science, technology, and capital could both devastate a resource and try valiantly to restore it. But when fishermen journeyed to the banks of the Merrimack, self-interest seemed to dampen their concern for the public good.

Eventually, even the prized effort to restock the river with salmon succumbed. Here the fault lay not with overfishing, but in the failure to keep the river continually free of obstruction. It was true that during the late 1880s and into the next decade, salmon ascended the river in numbers unheard of since the early part of the nineteenth century. According to one estimate, in the years from 1887 to 1891, a record six thousand salmon are believed to have made it to spawning grounds upstream.[136] But the years that followed saw the demise of the Merrimack's salmon runs. In the period from 1892 to 1895, construction of a dam at Sewall's Falls in Concord, New Hampshire, blocked off the river. Spanning five hundred feet in length, the dam eventually generated electricity at the falls.[137] Until a fishway was provided in 1895, migratory species of fish were unable to ascend any further upstream. When a fishway was built, imperfections in its design may have failed to allow salmon to pass.[138] Commenting on this untoward development, the Massachusetts Commission on Inland Fisheries and Game wrote: "When we recall the aggressive message of a former governor of New Hampshire, complaining that Massachusetts had deprived his

---

[135] Mass. Commission on Inland Fisheries, *Twelfth Annual Report* (1878), Public Doc. 34, 8.
[136] Stolte, *Forgotten Salmon of the Merrimack*, 108.
[137] The Concord Land and Water Power Company, which built the dam, is discussed in the Epilogue.
[138] Stolte, *Forgotten Salmon of the Merrimack*, 115, n. a, 120, n. b.

State of migratory fish by impassable dams at Lawrence and Lowell, we regret to find so little appreciation of the efforts of this State to restore the fish."[139]

History, some say, repeats itself. And it may well have seemed so in 1896 as the Lawrence fishway was swept off by a spring flood, delivering yet another blow to the river's migratory fish. In March 1896, large chunks of ice, trees, and logs came rushing downstream, carrying away the upper third of the fishway. According to one observer, salmon were seen that summer jumping up into the air at the foot of the dam.[140] Between 1892 and 1897, no salmon reached their spawning grounds.[141] Once again, the troublesome Lawrence Dam had upset the ecological order, dooming the prospects for restoring fish to the valley.

The industrial transformation of the Merrimack valley left an enduring ecological legacy. A period of over fifty years of river obstruction began in the eighteenth century with efforts to make the Merrimack navigable. More extensive obstruction followed with the subsequent rise of the Waltham–Lowell system. Many people welcomed these developments as part of the onward march of civilization and progress. But as economic growth worked its way up the valley, it created a rather different regional ecology, one that was difficult to reverse.

Nature is delicate and at times not terribly forgiving. While the extinction of living organisms has long been a part of history, reversing this process has proved to be no simple matter. The effort to restock the Merrimack River with fish confronted formidable obstacles: the enduring ecological impact of industrial change, the contradictions inherent in converting fish into a form of private property, and, not least of all, just plain, ordinary greed. It is easy to destroy but painfully difficult to restore; at least this is part of what the attempt to redevelop the Merrimack fisheries teaches.

---

[139] Mass. Commission on Inland Fisheries and Game, *Annual Report* (1895), Public Doc. 25, 5.

[140] Mass. Commission on Inland Fisheries and Game, *Annual Report* (1896), Public Doc. 25, 29–30.

[141] Stolte, *Forgotten Salmon of the Merrimack*, 121. In 1895, the Massachusetts legislature passed an act repealing restrictions on fishing for shad and alewives in the estuary, despite the objections of the state fish commission. See Act of 28 Feb. 1895, ch. 88, [1895] Mass. Acts & Resolves 83; Mass. Commission on Inland Fisheries and Game, *Annual Report* (1895), 7.

The effort to restock inland fisheries illustrates another important point. It is perhaps hard to see any common ground between the industrial development of water and the later move to reverse this process by restoring fish. Yet the two developments have a great deal knitting them together. Although anger at the lower Merrimack waterpower companies was partly responsible for the first calls for fish restoration, that hostility never extended to the companies' ethos of production. Likewise, proponents of fish culture demonstrated a deep commitment to the idea of maximum production. The attempt to reinvigorate the Merrimack valley fisheries was given legitimacy by appealing to the creed of production maximization. That ethos found support in the principle of private property, the belief that exclusive control over property created a stable legal context for commercial development. Having made known their preference for production, the Massachusetts fish commissioners perhaps were destined to embrace private property. Equipped with such assumptions, it is no wonder that they found manufacturing interests congenial to their endeavors.

Still, manufacturers only supported fish restoration so long as it did not clash with their own objectives for water. Fishways were fine unless they jeopardized the use of water for production. Employing water to manufacture commodities or to cultivate fish could both be justified by invoking the notion of maximum production. Yet somehow the attempts to promote fish culture seemed weaker and less persuasive than using water for manufacturing. Compared with the production of textiles, the prospect of cultivating fish was less compelling from an economic standpoint. The potential gains and benefits of using water for production were not all equal. Thus, the fish culture program struggled to convince New Englanders of the economic utility of the enterprise.

Finally, the history of the Merrimack fisheries demonstrates a major theme in this study. Nature incorporated into human agendas, manipulated and controlled for human ends, has been a persistent issue throughout these chapters. This dynamic is evident in one form or another from the very first attempts to use the river and its tributaries for boating, fishing, and energy. But the drive to triumph over nature became overwhelming in the later rise of the Waltham–Lowell system. That system had, at its heart, the control of water. Just as certainly, the attempt to restock the Merrimack and other inland waterways represented a variation on this same theme. It displayed the efforts of a region transformed ecologically by industrial change yet wedded to the same principles of control and mastery over nature. Indeed, the struggle to restock the region's waterscape with artificially propagated fish epitomizes the redesign of nature to suit the needs of an economic culture.

7

FOULED WATER

By the second half of the nineteenth century, they were everywhere. Perched over power canals, crammed in beside waterfalls, scattered along brooks, the Merrimack valley mills produced all sorts of things: textiles, timber, paper, leather, combs, chairs, iron tools, carriages, doors, match boxes, staves, tables, spools, pails, railway cars, pianos, buttons, scythes, hats, wagons, cigars, and mattresses. The mills needed workers, and thus where factories emerged, people followed – many of them. In 1870, Lowell's population topped 40,000, Lawrence had close to 29,000 people, and Manchester over 23,000.[1] Fitchburg and Clinton, Massachusetts, industrial towns in the Nashua River valley, each almost doubled in size between 1855 and 1875, their combined population reaching over 19,000 by the latter date.[2] Thousands of city dwellers, mills producing everything from the mundane to the grand – all in one way or another were dependent on the waters of the Merrimack valley.

The valley's vast waters were the source of much economic opportunity. They also promised, if handled wisely, to sustain human life. Unfortunately, what is readily available and abundant is often easily taken for granted and subject to neglect. River water is always on the move, making it a convenient, simple way to dispose of whatever waste people and industry generated. "The temptation to cast into the moving water every form of portable refuse and filth, to be borne out of sight, is too great to be resisted."[3] Those are the words of William R. Nichols and George Derby, the two men charged with investigating river pollution in 1873 for the Massachusetts State Board of

---

[1] U.S. Census of Population, 1870.
[2] Mass. Bureau of Statistics of Labor, *Census of the Commonwealth of Massachusetts 1905*, 4 vols. (Boston, 1908–10), 1:825, 834.
[3] William Ripley Nichols and George Derby, "Sewerage; Sewage; The Pollution of Streams; The Water-Supply of Towns," in MSBH, *Fourth Annual Report* (1873), 40.

Health (MSBH). The quality of New England's water had changed a great deal over the course of the nineteenth century. By the 1870s, some people were dipping into the waters of the region, not to use it for production, but to analyze it in laboratories to determine its pollution level.

The economic development of the Merrimack valley helped put an end to the region's fish runs, to the set of ecological relations that had long dominated the area. But industrial transformation did more than destroy. It also created a new ecology of its own with far-reaching effects on the water quality of the region's rivers, and ultimately on human existence itself. The growth of manufacturing and the rise of large cities placed tremendous demands on water. Rivers were used to generate energy for production, to carry off human and factory waste, and ultimately to supply cities with water for domestic use. That is a lot to ask of any resource, no matter how plentiful. The heavy demands placed on the Merrimack River watershed generated an increase in river pollution and the eventual spread of waterborne disease. A new set of environmental problems eventually emerged with widespread consequences for water and for the health of the people who lived in the valley.

It sounds almost fanciful, but under some circumstances it is true: Rivers have the unique capacity to purify themselves. Much of the waste disposed of in New England's waters over the centuries has been organic in nature. Organic pollution, substances ranging from soaps and animal oils to waxes and resins, consists of carbon compounds that can be broken down by microorganisms present in river water. For the biochemical reaction to work properly, dissolved oxygen is needed. If the amount of organic pollution is limited, and the stream has a sufficient level of dissolved oxygen, it is possible for aerobic bacteria to transform organic matter into fairly benign substances. The stream is thus said to have experienced *self-purification*. A number of factors can interfere with this process. An especially large quantity of organic pollution can deplete the stream's supply of dissolved oxygen. In addition, inorganic pollutants, such as acids and alkalis present in industrial waste, can destroy the bacteria needed to carry out self-purification. What remains of the organic matter is broken down by organisms called anaerobic bacteria that do not need free oxygen.[4]

An exact understanding of the self-purification of rivers – a more complicated process than the above suggests – was largely unknown until late in the nineteenth century. This, of course, did not stop industry from dumping waste into New England's rivers. The history of industrial pollution, its precise degree and extent, is difficult to determine. Still, the enormous amount of printed cotton textiles produced by the Waltham–Lowell factories must have resulted in substantial water pollution. At Lowell in 1835, 11.2 million yards of textiles were printed and dyed by the Merrimack and Hamilton companies, a figure that ten years later had grown slightly to 14.6 million. By 1860, however, the two companies produced over 29.2 million yards of dyed and printed goods; the Lowell Bleachery (in operation since 1847) was dying an additional 15 million yards and bleaching 8 million pounds per year.[5] The Pacific Mills in Lawrence at this same date annually printed 25 million yards of textiles.[6]

Organic pollution accounted for most of the waste produced in the process of dying and printing textiles.[7] Commonly used dyes included madder (a red dye obtained from the madder root), logwood (a black or brown dye from this tropical American tree), and peachwood, all of which added to the organic pollution load of the Merrimack River. Since the dying and printing of textiles involved the use of mordants for fixing color to textile material, other substances were needed in addition to dyes. These were largely inorganic chemicals, less easily tolerated by the Merrimack, and included sulfuric acid, muriatic acid, lime, and arsenate of soda (a compound containing the poisonous element arsenic). Only a small amount of the actual dye remained fixed to the cloth, while the bulk of it entered the river as waste, causing the water to be discolored.

[4] Louis Klein et al., *River Pollution*, 3 vols. (London, 1959–66), 2:35–9. Also see the discussion in R. A. Howe, *The Chemistry of Our Environment* (New York, 1978), 272–85.

[5] Figures are derived from "Statistics of Lowell Manufactures, January 1, 1835" (handwritten broadside, 1835), MATH; *Statistics of Lowell Manufactures, January 1, 1845* (Lowell, 1845); *Statistics of Lowell Manufactures, January 1, 1860* (n.p., 1860). These figures were given as weekly totals and were multiplied by 51 to give the approximate output for the year.

[6] *Statistics of Lawrence, (Mass.,) Manufactures 1860* (Manchester, N.H., 1860).

[7] The following discussion of industrial pollution is based on James P. Kirkwood, "The Pollution of Rivers; An Examination of the Water-Basins of the Blackstone, Charles, Taunton, Neponset, and Chicopee Rivers; With General Observations on Water-Supplies and Sewerage," in MSBH, *Seventh Annual Report* (1876), 37–57.

Table 7.1. *Number of waterpowered mills[a] and workers: Fitchburg, Clinton, and Leominster, Massachusetts, 1850 and 1876*

City[b]	1850		1876	
	Mills	Workers	Mills	Workers
Fitchburg (12,289)	32[c]	610	27	968
Clinton (6,781)	8	1,094	5	1,953
Leominster (5,201)	33	396	14	411
Total	73	2,100	46	3,332

[a]Figures do not include sawmills.
[b]Figures in parentheses equal the total population in 1875.
[c]There was one additional waterpowered mill that was illegible.
Sources: Manuscript Schedules, U.S. Census of Manufacturing, 1850; MSBH, *Eighth Annual Report* (1877), 25–34.

The washing of newly printed cloth at Lawrence caused "a violent rush from the race-ways of inky liquid, which . . . may often be distinctly traced three hundred or four hundred feet across the stream."[8]

The Waltham–Lowell mills were, of course, not solely responsible for all the pollution found in the Merrimack River. Much of the industrial waste originated on the Merrimack's tributaries, especially along the Nashua River. By midcentury, as the Waltham–Lowell mills settled in along the Merrimack, industrial activity in the Nashua River valley proceeded apace. Table 7.1 reveals seventy-three mills at midcentury in the valley's three largest towns (not including Nashua, New Hampshire), Fitchburg, Clinton, and Leominster, Massachusetts. A quarter of a century later, only forty-six mills remained. Yet the laborers involved in manufacturing rose sharply by almost 60 percent from 2,100 to 3,332 workers, suggesting more production and probably more pollution.

More important, a shift toward industries that tended to pollute more occurred in the valley (Table 7.2). Paper mills are notorious for their pollution. Transforming old rags into paper

[8] Mass. Legislature, Commission on Obstructions to the Passage of Fish in the Connecticut and Merrimack Rivers, "Report of the Commissioners" (Jan. 1866), S. Doc. 8, 14.

Table 7.2. *Types of polluting industry: Fitchburg, Clinton, and Leominster, Massachusetts, 1850 and 1876*

Industry	1850		1876	
	Mills	Workers	Mills	Workers
Paper mills/wood pulp	7	114	11	271
Cotton mills	5	775	4	1,400
Woolen mills	3	101	5	290
Tanneries/leather	—	—	3	119

Sources: Manuscript Schedules, U.S. Census of Manufacturing, 1850; MSBH, *Eighth Annual Report* (1877), 25–34.

pulp, the first principal task in papermaking involved breaking down the material by boiling it in a caustic alkali solution, creating wastewater generally passed into streams. The pulp was then bleached using a solution of lime chloride and sulfuric acid. Fitchburg had five paper factories in 1850 employing seventy-two workers. Twenty-six years later, there were double this number of paper and pulp mills, with three times as many laborers.[9] Likewise, significant pollution resulted from the manufacture of woolen textiles, a complicated process with a number of steps requiring water. Wool had to be washed and boiled in a solution of detergents (including soda ash and phosphate of soda). The wastewater from this procedure was flushed into streams, as were wastes from the dying of the woolen yarn. The number of woolen mills in the towns of Fitchburg, Clinton, and Leominster rose from three to five between 1850 and 1876, while the work force almost tripled in size. Finally, tanneries caused serious water pollution. The preparation of animal hides produced large quantities of suspended matter in water, including particles of hair, flesh, and lime, all of which put heavy stress on a river's supply of dissolved oxygen. While Leominster had no tanneries at midcentury, twenty-six years later there were three leather manufacturers with 119 workers.[10]

In 1876, the MSBH surveyed water pollution in the Nashua River basin. At that time, ninety-two mills with over five thou-

---

[9] Manuscript Schedules, U.S. Census of Industry, Fitchburg, Mass., 1850; Charles F. Folsom, "The Pollution of Streams, Disposal of Sewage, Etc.," in MSBH, *Eighth Annual Report* (1877), 27–9.

[10] Folsom, "Pollution of Streams," 28.

sand workers dwelled beside the valley's waters. Cotton mills, paper mills, and woolen mills, in that order, were the three leading industries, together accounting for over 60 percent of the mills in the watershed. The survey noted that none of the valley towns as yet had comprehensive sewage systems that drained into either the Nashua River or its tributaries. Thus the primary source of water pollution came from industrial activity.[11] Parts of the valley were evidently free from pollution, but at other points serious amounts of industrial waste fouled the river. As part of its investigation, the board asked local selectmen for information on the extent of pollution in their towns. The north branch of the Nashua in Fitchburg was described as follows:

> The water of the Nashua, in passing this city, is extensively polluted by the wash of nine paper mills, four woolen mills, two cotton mills, gas works, and other manufacturing establishments on or in the immediate vicinity of the stream. The water presents a dirty appearance, and people generally shrink from bathing in it. It receives the whole sewage of the city, so far as sewage is disposed of at all. . . . All the chemicals employed in paper mills and different manufacturing establishments – excrement, dyestuffs, etc., and street washings – find their way directly into the stream. The extent of the pollution is great. The color of the water is materially changed in its whole course through the city during low water.[12]

The abject pollution of the Nashua River at Fitchburg stopped some manufacturers from using water for industrial purposes (apart from as a source of waterpower). One paper mill filtered the water before using it, while a woolen mill could only use the water for dying dark colors.[13]

Despite this evidence, the report found the valley free from "marked offensiveness in many places." In part, this was because the watershed in the 1870s was still predominantly rural. A single woolen mill perched along Britain's River Aire – using 320,000 pounds of logwood and other dyes plus 700,000 pounds of soap per year – employed nearly three-quarters of all the operatives working in the Nashua River valley in Massachusetts (part of the valley lay in New Hampshire).[14]

Still, enough pollution existed in some places to warrant the MSBH's attention. In the town of Shirley, sawdust, which had

[11] Ibid., 24, 35.
[12] Ibid., 43.
[13] Ibid., 32.
[14] Ibid., 46.

accumulated in Mulplus Brook, destroyed the local supply of brook trout.[15] A brook running through North Leominster was so contaminated with tannery waste that residents were compelled "to shut their windows at times to keep out of their houses the vile odors."[16] And as the Nashua River headed northeasterly, crossing the state line into New Hampshire, the city of Nashua added further pollution to the river. That city's waste was the direct responsibility of the neighboring state. But since the Nashua's water flowed into the Merrimack and on back into Massachusetts, the water pollution eventually became a concern for the MSBH. With over ten thousand people in 1870 and host to Waltham–Lowell-style mills, two dye works, and one paper mill, the city of Nashua, New Hampshire, presented a significant pollution threat. According to one report, below the mills on the river "the water is almost black, is unfit to drink, and in very hot weather is sometimes odorous, but not to the extent of causing general complaint."[17]

With the exception of its headwaters, the MSBH found the water in the Nashua River unfit for drinking. River pollution, in the board's opinion, also contributed in part to the decline of the Nashua's fisheries. Factory waste from the town of Clinton was reported to have killed fish two miles downstream in the river.[18] Still, whatever the ecological degradation, to clean up the river, the board concluded, would be a rather formidable task, not worth the money and trouble. Some amount of pollution was in their view unavoidable, an inevitable product of urban, industrial life. "Until we have better means of disposal of our refuse than at present," the board wrote, "some of our rivers must be used, more or less, to scavenge the country."[19]

Nothing compares with convenience for frustrating change. Rivers were a handy, expedient way for disposing of New England's collective waste, to see it borne off downstream by the current, no sooner out of sight than out of mind. But years and years of abuse and neglect had by the 1870s left a rather disturbing mark on the waterscape. In that decade, as concern emerged over public health in Massachusetts, some state offi-

[15] Ibid., 45.
[16] Ibid., 46.
[17] The statement was made by the board of health of the city of Nashua and is quoted in ibid., 67.
[18] Ibid., 61–2.
[19] Ibid., 62–3.

cials began to wonder about the costs of water pollution. As George Derby, secretary of the MSBH put it, "There will never be wanting advocates of any application of natural forces which may lead to individual or corporate profit, while considerations of public health are always less obvious."[20]

A number of extensive river surveys were done by the MSBH in the 1870s. Under the guidance of James Kirkwood, a civil engineer from Brooklyn, New York, the MSBH assessed the public health threat posed by water pollution along the Blackstone, Charles, Taunton, Neponset, and Chicopee rivers.[21] By the latter part of the nineteenth century, the rivers of southern New England were packed with factories. Forty-four woolen mills and 27 cotton textile factories crowded into the 312 square miles of the Blackstone River valley that lay in Massachusetts. There was 1 mill for about every 2 square miles of land in the Neponset River valley (116 square miles with 52 mills), producing textiles, leather, tools, and chemicals. Iron works, 30 in all, piled into the Taunton River valley. And spread out across the more substantial expanse of the Chicopee River valley (718 square miles) were 125 factories filled with the busy sounds of close to nine thousand workers.[22]

The Charles River valley, host to the incipient Waltham–Lowell system early in the century, had changed a great deal since then. When the Boston Manufacturing Company in Waltham prepared for business in 1812, twenty-three mills – seven of which were saw, grist, and fulling mills – dotted the shores of the Charles River.[23] By 1875, that number had increased somewhat to thirty. More important, the types of industrial activity changed, much as they did in the Nashua River basin. The number of woolen mills along the river had gone from zero to four between the two dates. Already in 1812, there were five paper mills, a figure that nearly doubled to nine by 1875.[24]

---

20 George Derby, "Mill-Dams and Other Water Obstructions," in MSBH, *Third Annual Report* (1872), 60.
21 Kirkwood, "Pollution of Rivers," 23–174. Later in the decade, surveys were done for the Hoosac and Housatonic rivers. See "Drainage and Health; Sewerage; and the Pollution of Streams; Including a Draft of a Law," in MSBH, *Ninth Annual Report* (1878), 3–80.
22 Kirkwood, "Pollution of Rivers," 82, 94, 136, 118. Statistics on the size of the actual watersheds are from the maps included with the study.
23 Edward Pierce Hamilton, "Early Industry of the Neponset and the Charles," *Proceedings of the Massachusetts Historical Society* 71 (1953–7): 111.
24 Kirkwood, "Pollution of Rivers," 98–9.

Despite the expansion of industry, Kirkwood and his associates found the water in most of the river fit for domestic consumption if filtered properly. Still, they compared the river and its watershed of forty-odd mills and their trail of polluting waste to the Thames River above London.[25] By the late nineteenth century, the Thames and the rivers of northern England were among the foulest in the world.[26]

Unfortunately, no extensive survey was done for the Merrimack River itself in 1875. But the chemical composition of the Merrimack's water was analyzed by William Ripley Nichols, a chemist at the Massachusetts Institute of Technology. Regarded as an expert in applying chemistry to public health issues, Nichols was well known for his work on water pollution.[27] On a cold afternoon in November 1872, Nichols and George Derby visited the Merrimack and Concord rivers near Lowell. They collected water samples, which they brought to a laboratory for tests and found free, in their opinion, of serious pollution. Yet they did not recommend the Merrimack's water below Lowell as a source of drinking water.[28] Nichols returned to the Merrimack the following year and noted the great number of factories and cities that lined the river's shore. More samples were taken including one below the city of Lawrence. "At this point," he wrote, "the surface of the river is dotted over with bits of wool and cotton; the water is covered with a greasy film, and sometimes thickly strewn with flocks of soap-suds."[29]

Nichols knew that the Merrimack Manufacturing Company upstream at Lowell alone consumed 225,000 pounds of starch, 2.5 million pounds of madder, 1.1 million pounds of sulfuric acid, and 350,000 pounds of soda ash, much of which went flowing into the river. "From what has been said of the magnitude of the manufacturing operations carried on upon the banks of the river," explained Nichols, "we might at first sight suppose that the effect would be to make the water very foul,

25 Ibid., 108.

26 Christopher Stone Hamlin, "What Becomes of Pollution? Adversary Science and the Controversy on the Self-Purification of Rivers in Britain, 1850–1900" (Ph.D. diss., University of Wisconsin – Madison, 1982), 9, 14–15.

27 For a brief biographical entry on Nichols, see James Grant Wilson and John Fiske, eds., *Appletons' Cyclopaedia of American Biography,* 7 vols. (New York, 1888–1901; reprint, Detroit, 1968), 4:514.

28 Nichols and Derby, "Sewerage," 89–91.

29 William Ripley Nichols, "On the Present Condition of Certain Rivers of Massachusetts," in MSBH, *Fifth Annual Report* (1874), 65, 69.

and to cause it to be unfit to use for any domestic purpose."[30] Not so, he argued, as his own chemical analysis confirmed. Although the river appeared filthy below Lawrence, the mean of eleven water samples taken there showed a small amount of total solid matter (4.43 parts per 100,000, while above the city the figure was just 4.10).[31] What did Nichols think happened to the pollutants discarded into the Merrimack River?

Three factors, in his opinion, helped purify the river: oxidation, deposition, and dilution.[32] Oxidation of organic refuse was, to his mind, the least important of the three processes. Although it is now known that bacteria present in rivers help oxidize organic matter, breaking it down into benign forms, this was not a conclusion arrived at easily. At the time that Nichols lived, a great deal of contention – most of it centered in Britain – surrounded the notion that a stream could engage in self-purification. During the 1860s, Edward Frankland, a British expert on water supply, conducted experiments on the scientific validity of river self-purification. His results cast serious doubts on the theory behind this process. Whatever oxidation of organic matter occurred in rivers happened slowly. Writing in 1868, he remarked, "There is no river in the United Kingdom long enough to effect the destruction of sewage by oxidation."[33]

Frankland's experimental work with water served as a model for Nichols.[34] More important, his low opinion of the capacity of a river to oxidize organic matter seemed to inform Nichols's view as well. Nichols found a decline in the oxygen present in the water as it flowed through the city of Lowell, but he denied the significance of this decline. A large number of observations over time, he believed, might indicate a decline in oxygen in water below manufacturing cities. Yet he chose not to investigate this hypothesis experimentally.[35] Nichols was, however, a bit more hopeful about the positive effects of deposition. Some of the wastes allowed to run into rivers were either deposited along the streambed or washed up along the shore. During freshets, these wastes were stirred up and carried out to sea.[36]

---

30 Ibid.
31 Ibid., 76.
32 Ibid., 76–82.
33 Quoted in Hamlin, "What Becomes of Pollution?" 376. Also see Hamlin's discussion of Frankland and the theory of self-purification, 297–300, 369–77.
34 Ibid., 376–7.
35 Nichols, "Present Condition of Certain Rivers," 77.
36 Ibid., 77–8.

Purification of the Merrimack, in Nichols's opinion, was mostly due to the simple process of dilution. Enormous quantities of waste went into the river, but these were small compared to the Merrimack's volume of water. Again, Nichols turned to the water samples taken above and below the city of Lawrence where large amounts of pollution, both manufacturing waste and sewage, entered the river. As indicated above, he observed but a slight increase in the total amount of solid matter. There was, however, a decline in the level of chlorine. Chlorine was found both in sewage and in substances used by manufacturers and disposed of in rivers. Since no natural processes can remove chlorine from streams, the decline, Nichols deduced, could only result from dilution.[37] Assuming 2,100 cfs as the Merrimack's average summer flow at Lowell, he calculated the need for 100 tons of dry waste to increase the quantity of solid matter in solution by as little as one grain to the gallon. The mean flow of the river, he believed, was significantly above that figure (5,400 cfs), making "dilution alone . . . sufficient to account for the slight apparent effect of the discharge into the stream of much that is offensive and noxious."[38]

After his work on the Merrimack, Nichols completed chemical assessments of the Blackstone, Charles, Neponset, Chicopee, and Taunton rivers as part of an MSBH water pollution survey. Except for the Blackstone River downstream from Worcester, those results showed, in the words of the final report, "that the condition of the five rivers is not yet very bad."[39] When the Nashua River was surveyed the following year, Nichols was again called on to perform a chemical analysis of the water. Interpreting the results, the MSBH found "a decided, but not excessive, increase in the amount of pollution, as the outlet is approached." Still, the board cautioned that the most serious source of river pollution came from human waste found, as yet, in relatively small quantities in the Nashua River.[40]

In less than a century, the rivers of southern New England had been eagerly driven down an industrial path. Their value as an energy source and a means of waste disposal combined to draw factories to their shores. Mills filled in along the banks, and the rivers themselves filled up with the collective waste of

37 Ibid., 78–9; Hamlin, "What Becomes of Pollution?" 412–13.
38 Nichols, "Present Condition of Certain Rivers," 80–1.
39 MSBH, *Seventh Annual Report* (1876), 9.
40 Folsom, "Pollution of Streams," 59.

Table 7.3. *Population of major Merrimack valley cities, 1850–80*

City	1850	1860	1870	1880
Concord, N.H.	8,576	10,896	12,241	13,843
Fitchburg, Mass.	5,120	7,805	11,260	12,429
Haverhill, Mass.	7,205	11,683	15,106	21,115
Lawrence, Mass.	8,282	17,639	28,921	39,151
Lowell, Mass.	33,383	36,827	40,928	59,475
Manchester, N.H.	13,932	20,107	23,536	32,630
Nashua, N.H.	5,820	10,065	10,543	13,397
Newburyport, Mass.	9,572	13,401	12,595	13,538

Sources: Massachusetts Bureau of Statistics of Labor, *Census of the Commonwealth of Massachusetts 1905*, 4 vols. (Boston, 1908–10), 1:834, 841–2, 847, 849–50, 860; U.S. Census of Population, 1850, 1860, 1870, 1880.

Table 7.4. *Percent population change in Merrimack valley cities, 1850–80*

City	1850–60	1860–70	1870–80
Concord, N.H.	27.05	12.34	13.09
Fitchburg, Mass.	52.44	44.27	10.38
Haverhill, Mass.	62.15	29.30	39.78
Lawrence, Mass.	112.98	63.96	35.37
Lowell, Mass.	10.32	11.14	45.32
Manchester, N.H.	44.32	17.05	38.64
Nashua, N.H.	72.94[a]	4.75	27.07
Newburyport, Mass.	40.00	−6.01	7.49

[a]Represents the annexation of the village of Nashville with the incorporation of the city of Nashua in 1853.

manufacturing, with the dregs of industrial production. By the 1870s, the ecological impact of this process had become terribly apparent. Worse still, there were other threats to rivers beyond those posed by the factories that closed in upon the water.

As the Waltham–Lowell mills gathered up water for production, they also collected large numbers of people in the process. By 1860, the Merrimack valley's largest cities were Lowell, Manchester, and Lawrence (Table 7.3). The city of Lawrence had more than doubled in size during the decade of the 1850s. During the second half of the nineteenth century, the Merrimack valley cities continued to grow at rates that seldom fell

Table 7.5. *Approximate population densities of Merrimack valley cities, 1850–80[a]*

City (area)	1850	1860	1870	1880
Concord, N.H. (63.0 sq. mi.)	136	173	194	220
Lawrence, Mass. (7.2)[b]	1,150	2,450	4,017	5,438
Lowell, Mass. (14.3)[c]	2,334	2,575	2,862	4,159
Manchester, N.H. (35.0)	398	574	672	932
Nashua, N.H. (30.0)	—	336	351	447

[a]Figures are given in persons per square mile.
[b]Part of Methuen was annexed in 1854; parts of Andover and North Andover were annexed in 1879.
[c]Part of Dracut was annexed in 1851. Portions of the towns of Chelmsford and Dracut were annexed to the city in 1874 and 1879; sections of Tewksbury were added in 1888 and 1906.
Sources: Mass. Department of Public Health, *Sources of Pollution: Merrimack River Valley* (Boston, 1938), 10–12; U.S. Census of Population, 1850, 1860, 1870, 1880.

below double digits for any of the decades (Table 7.4). In 1880, the population of Lowell alone reached nearly sixty thousand, an increase of over 45 percent in ten years. At this date, Lowell, Lawrence, and Manchester were still the leading population centers, while the remaining major cities each had over twelve thousand people.

By the 1870s, eight cities sprawled out across the valley, all except for one (Fitchburg) straddling the shores of the Merrimack River. Along the lower Merrimack, population densities were especially high (Table 7.5 and Figures 7.1 and 7.2). Lawrence had roughly 70 percent as many people as Lowell in 1870, but only about half as much land. With population densities of over 2,500 people per square mile after 1860 – and in 1880 the population density of Lawrence climbed to over 5,000 per square mile – the textile cities along the lower Merrimack were swollen with people. The dense populations weighed heavily on the waterscape.

It is tempting to believe that cities are divorced from the natural world. Cities, with all their artificial contrivances, their imposing built environments looming over the horizon, seem

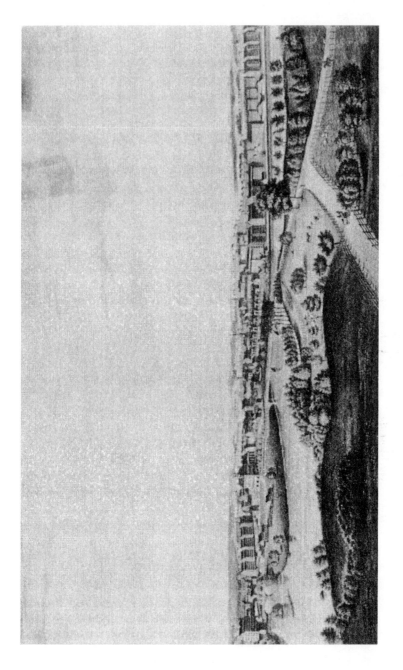

Figure 7.1. *View of Lowell, Mass.* Lithograph by E. A. Farrar, 1834. (Courtesy of the Museum of American Textile History.)

Figure 7.2. The same view of Lowell as in Figure 7.1 fifty years later. (From Frank P. Hill, *Lowell Illustrated* [Lowell, 1884]. Courtesy of the Museum of American Textile History.)

liberated from the constraints of nature. In fact, this is not the case. Urban agglomerations, however powerful as monuments to human ingenuity, remain as dependent on the natural world as the smallest rural community imaginable, stuck out there in the wilderness. The natural environment may well be more removed from the sight and consciousness of the urban dweller, but this in no way implies deliverance from the strictures of nature.

Cities, it has been said, have a metabolism all their own.[41] Collectively, they swallow up huge quantities of resources, then turn around and spew out the waste and whatever else they no longer need. To the urban manufacturer, water met the need for energy and a cheap source of waste disposal. For the huge number of people living in the Merrimack valley cities, water represented something more essential. Water is among the most basic of needs; indeed, human existence depends on it. The emergence of large urban centers like the ones that dominated sections of the Merrimack valley made the provision of water, and the process of getting rid of it, a complicated task. Satisfying the water needs of great numbers of people required a rational plan for controlling water in the interests of human ecology.

At first, there were no such plans. For half a century after its founding in the 1820s, Lowell depended on separate wells located throughout the city for its domestic supply of water.[42] Likewise, until the 1870s, the city of Manchester, New Hampshire, relied on wells and springs. Spring water was in especially ample supply. As one observer recalled, "Manchester might well have been called at one time, the city of springs, for turn which way you might, forty years ago, the thirsty man could have found clear, sparkling water gushing forth from the hills and plains."[43] On occasion, small companies were formed to supply water. The Concord Aqueduct Association began in 1829 to bring water from a spring to this New Hampshire city. Otherwise, wells met the bulk of the city's water needs.[44] Tap-

[41] Abel Wolman, "The Metabolism of Cities," in *Cities: A Scientific American Book* (New York, 1969), 156–74.

[42] MSBH, "An Inquiry into the Causes of Typhoid Fever, as it Occurs in Massachusetts," in MSBH, *Second Annual Report* (1871), 139–40.

[43] William B. Blake, "The Water Supply of Manchester," *Manchester Historic Association Collections* 3 (1902–3): 80.

[44] James O. Lyford, ed., *History of Concord New Hampshire*, 2 vols. (Concord, N.H., 1903), 1:527.

ping wells, springs, ponds, and rainwater cisterns, the people of the Merrimack valley obtained their water for most of the nineteenth century on an ad hoc basis.

With no large-scale public waterworks before the latter part of the century, the quantity of wastewater remained relatively low. Little need existed for a comprehensive system of sanitary sewers. As late as the early 1870s, Lowell still had no general sewage system.[45] Cesspools and privies were used to handle human bodily waste, the water left to gradually seep out into the surrounding soil. Still, rivers were sometimes relied on to dispose of wastewater. The massive Lancaster Cotton Mills in Clinton, on the south branch of the Nashua River, had 1,200 operatives by the 1870s using privies that discharged directly into the stream.[46] The Essex Company in Lawrence drew up its plans for the city to allow for drainage and an underground sewage system. By 1850, three main sewers – one large enough for a person to walk through – discharged wastewater into the Merrimack and Spicket rivers.[47] The vast majority of human waste, however, found its way onto the land. In 1876, the MSBH surveyed nineteen cities and towns in the Merrimack valley and found all but six (Ashburnham, Clinton, Lancaster, Haverhill, Lowell, and Lawrence) disposing of half or more of their sewage on the surface.[48] The land, as historian Joel Tarr has remarked, remained the principal sink for human waste.[49]

By the 1860s, the need for reliable supplies of water and comprehensive sewage systems became more pressing. According to L. Frederick Rice, the civil engineer who planned the public water supply and sewers of the city of Lawrence, privies and cesspools worked in sparsely settled areas. But this method of disposal could not persist forever:

> As the density of population increases and the saturation of the soil becomes more complete, the presence of this mass of decomposing matter begins to exercise an injurious effect upon the health of the community. When the people derive the water used for drinking, cooking and other domestic purposes from wells excavated in the

---

[45] "Inaugural Address of His Honor Edward F. Sherman, Mayor of the City of Lowell, January 2, 1871," in *CDL*, 1870–1 (Lowell, 1871), 25.

[46] Folsom, "Pollution of Streams," 39.

[47] Mass. Sanitary Commission, *Sanitary Survey of the Town of Lawrence* (Boston, 1850), 4.

[48] Frederick Winsor, "The Water-Supply, Drainage and Sewerage of the State, From the Sanitary Point of View," in MSBH, *Seventh Annual Report* (1876), 202–5.

[49] Joel A. Tarr, "The Search for the Ultimate Sink: Urban Air, Land, and Water Pollution in Historical Perspective," *Records of the Columbia Historical Society of Washington, D.C.* 51 (1984): 3.

soil thus more or less saturated with filth, this water cannot long escape contamination.[50]

In fact, water supplies could become contaminated even in relatively unpopulated regions. A doctor visiting a farmhouse outside the city of Concord, New Hampshire, noticed a woman in the kitchen straining water she had just pumped out of a well. When he inquired into her purposes he was told "'that it was done to strain out the dead potato-bugs that had washed into the well during the spring rains.'" The insecticide Paris green had been used the previous year on a potato field nearby the well.[51]

This incident took place beyond the limits of the city. If anything, however, the threats posed to well water by dense urban populations were even more serious. In Manchester, a report written in 1872 pointed out that "shallow wells from which so many citizens are now supplied with water, are becoming more and more foul each season . . . [while] the rain-fall by which they are supplied is contaminated by the filth and rubbish in the alleys and excrements [sic] in the streets."[52] Responding to the need for pure water – made more pressing by the demand for a reliable supply to fight fires – the cities of Concord, Manchester, Lowell, and Lawrence all constructed municipal waterworks during the 1870s.[53]

Along the central and upper parts of the Merrimack River, cities turned to nearby ponds and brooks for water. Beginning in the 1850s, the water supply of the city of Nashua came from Pennichuck Brook. The brook eventually flowed into the Merrimack and drained, with its tributaries, roughly twenty-five square miles of land. In 1854 and again in 1869, dams built across the stream were used to impound water in reservoirs.[54] Concord, New Hampshire, brought water three miles by pipe into the city from Long Pond, an ample body of water three

[50] L. Frederick Rice, *Report on a General System of Sewerage for the City of Lawrence* (Lawrence, 1876), 2–3.

[51] A. H. Crosby, "Water Pollution – Public and Private," in N. H. State Board of Health, *First Annual Report* (1882), 250–1.

[52] J. T. Fanning, "Report on Sources of Water Supply, for the City of Manchester, N.H., With Estimates of Cost," in Board of Water Commissioners of the City of Manchester, N.H., *First Annual Report* (1873), 17–18.

[53] See, e.g., Lyford, *History of Concord*, 1:527.

[54] Arthur B. Graves, comp., "A Brief History of the Pennichuck Water Works of Nashua, New Hampshire" (ms., 1923), Hunt Room, Nashua Public Library, Nashua, N.H., see foreword, chs. 1, 11.

hundred acres in extent with a clean, sandy bottom. By 1873, the city's waterworks was in operation after over 92,000 feet of pipe were laid for distributing the water.[55]

In the 1870s, the city of Manchester began its search for a source of pure water. The well-watered terrain afforded a number of possibilities, including the suggestion that the Merrimack River itself be used. In making his report to the water commissioners, the city's chief engineer, J. T. Fanning, noted the Merrimack's plentiful waterpower. "It is this power," he wrote, "that has caused the river bank along its sides to be transformed, within forty years, from a wilderness of forest to a pleasant city with nearly thirty thousand souls."[56] But Fanning did not recommend the river as a source of water for the city, implying that such a plan would conflict with the use of water for industrial purposes. Instead, he suggested Lake Massabesic, 2,350 acres of water about four miles east of the city. That water, he pointed out, was free of sewage and industrial pollution.[57] When the system was finished in 1874, Fanning estimated that an average of 8 million gallons of water per day would be supplied. "After scanning the sources of city water supplies on all sides," wrote Fanning in delight over the completion of the project, "Manchester may still thankfully congratulate herself on the facilities which Nature placed within her reach."[58]

On the lower Merrimack, the cities of Lowell and Lawrence were considerably less fortunate. No sufficiently large and pure lakes for domestic supply existed near these cities, and they both had to turn to the Merrimack River. The city of Lowell had contemplated using the Merrimack's water for domestic consumption as early as 1838 and again in 1848 and 1860.[59] But after the Civil War, plans for creating a municipal water supply in the city moved ahead rapidly. A drought in the summer of 1865 caused further depletion of the city's wells, which were already of questionable purity. The introduction of water from the Merrimack River, in the opinion of L. Frederick Rice,

55 Board of Water Commissioners of the City of Concord, N.H., *Second Annual Report* (1873), 3, 11.
56 Fanning, "Report on Sources of Water Supply," 34.
57 Ibid., 34–9.
58 J. T. Fanning, "Engineer's Report," in Board of Water Commissioners of the City of Manchester, N.H., *Third Annual Report* (1874), 9–10.
59 *The Inaugural Address of His Honor Josiah G. Peabody, Mayor of the City of Lowell* (Lowell, 1866), 19–20.

would reduce the losses resulting from fires and contribute to a "great improvement in the health of the laboring classes."[60] In 1873, the Lowell Waterworks was completed, including a filtering gallery for use when freshets filled the water with large amounts of suspended matter.

Three years later, the city of Lawrence brought a similar waterworks into operation. Rice, who again served as the consulting engineer, found the city's wells unable to meet the growing demand for water. In recommending the Merrimack as the best possible source for pure water, Rice noted the "considerable objectionable matter" carried into the Merrimack by the Concord River. He also pointed out that a large number of factories on the Merrimack at Concord, Manchester, Nashua, and Lowell "discharge great quantities of impurity into the river, and on that account many persons have compunctions against using the water for domestic purposes."[61] Rice believed differently. In his view "one of the most remarkable qualities of running water is that of self-purification." Taking a view disputed by some scientists in England, Rice insisted "that when the most noxious matter has been thrown into a running stream, all traces of it have disappeared in the course of a few miles." The river's aquatic vegetation and animal life, he believed, removed some of the pollution. But "chemical action," in his opinion, played the key role in eliminating waste dumped into the river.[62]

By 1879, the valley cities of Lowell, Lawrence, Manchester, Concord, and Fitchburg had installed over 180 miles of water pipes.[63] The consumption of water rose dramatically. In Lowell, a daily average of close to 500,000 gallons of water was used in 1873, the first year the waterworks operated.[64] At the end of the decade, that figure had grown sharply by close to five times to 2.4 million gallons.[65] As had happened in most major American cities during the nineteenth century, the ad-

[60] L. Frederick Rice, "Report Upon the Proposed Lowell Water Works, With Estimates of the Cost of Construction," in *CDL*, 1866–7 (Lowell, 1867), 52–3.

[61] L. Frederick Rice, "Engineer's Report," *Report on Proposed Water Works at Lawrence, Mass.* (Lawrence, 1872), 25–6.

[62] Ibid., 26.

[63] Figures for all five cities are from Water Board of the City of Lowell, *Seventh Annual Report* (1880), in *CDL*, 1879–80 (Lowell, 1880), 29.

[64] Figure is derived from the monthly figures given in Water Board of the City of Lowell, *First Annual Report* (1874), in *CDL*, 1873–4 (Lowell, 1874), 31.

[65] Water Board of the City of Lowell, *Eighth Annual Report* (1880–1), in *CDL*, 1880–1 (Lowell, 1881), 20.

vent of piped-in water led to the breakdown of the former system for handling wastewater. Privies and cesspools could no longer adequately accommodate the growing quantity of polluted water.[66]

The vast increase in water forced the Merrimack valley cities to adopt systematic sewage systems. Again, the Merrimack River served as a convenient depository for the huge quantities of liquid waste created by the introduction of municipal water supplies. Both Lowell and Lawrence made plans for sewers in the same years their water systems came on line (1873 and 1876, respectively).[67] There were roughly thirteen miles of sewers in Lowell in 1873; it was estimated that close to three times this length would be necessary to meet the city's expanding need for wastewater removal.[68] Manchester commissioned a report on a system of drainage for the city in 1855. The report remained unpublished for eight years, suggesting little interest in its findings.[69] But after the city began taking water from Lake Massabesic, construction of a sewage system took place in the 1870s and 1880s. Forty miles of sewers crisscrossed the city by the end of 1890.[70]

Extensive water systems shifted the burden of pollution from the land to the Merrimack River. Where liquid waste once seeped from privies and cesspools into the surrounding ground, now urban wastewater passed directly into the Merrimack. With water being withdrawn for domestic use from the same river used for dumping sewage, the potential was high for contamination and the spread of water-borne disease. The city of Lawrence was especially vulnerable. That city was located but ten river

---

[66] For a general discussion of this problem as it affected U.S. cities during the nineteenth century, see Joel A. Tarr, "The Separate vs. Combined Sewer Problem: A Case Study in Urban Technology Design Choice," *Journal of Urban History* 5 (1979): 309–13.

[67] "Report of the Engineer, to the Committee on Sewers and Drains of the City of Lowell, on a General System of Sewerage," in *CDL*, 1873–4 (Lowell, 1874); Rice, *Report on a General System of Sewerage*.

[68] "Report of the Engineer, to the Committee on Sewers and Drains of the City of Lowell," 50.

[69] "Inaugural Address of the Hon. James A. Weston, Mayor," in *Twenty-Second Annual Report of the Receipts and Expenditures of the City of Manchester, For the Fiscal Year Ending December 31, 1867* (Manchester, N.H., 1868), 12–13; also see the actual report, *Slade's Report on the Drainage of the City of Manchester With the Grades of the Streets* (Manchester, N.H., 1863).

[70] "Report of the City Engineer," in *Forty-Fifth Annual Report of the Receipts and Expenditures of the City of Manchester, for the Fiscal Year Ending December 31, 1890* (Manchester, N.H., 1891), 64.

miles downstream from Lowell, which now discharged industrial and human waste into the Merrimack.

To cope with the threat of tainted water, both Lowell and Lawrence constructed filtering galleries. The one designed for Lowell was 1,300 feet long and drew the water through a gravel bed to filter out impurities. But the filtering gallery was mainly for use during spring freshets when the Merrimack's water contained large amounts of suspended matter.[71] The city's water supply was at first constantly filtered through the gallery, although it was not designed for year-round use. Then in 1874, the demand for water outstripped the capacity of the city's filtering scheme.[72] One critic of Lowell's water supply thought the filtering gallery was useless. In the late 1880s, he observed "dead and diseased cats and dogs, and other things of that sort that get into a public stream" passing unfiltered into the city's water.[73]

The threat sewage-polluted water posed to public water supplies went largely unaddressed. At one point someone proposed that sewage from the city of Lawrence be used for agricultural purposes, instead of letting it run into the Merrimack. But Frederick Rice rejected the idea in 1876 when making his recommendations for the city's sewage system. In Lawrence, he pointed out, "where the Merrimack river readily conveys away any noxious matter discharged into it, without injury to any other community, it is probably not advisable to make any attempt at sewerage utilization."[74]

To be sure, the river swiftly removed urban waste. But on one point, at least, Rice seemed mistaken: Casting wastewater into the river did indeed threaten other communities. Writing in the same year as Rice, an observer for the MSBH noted:

> The Merrimac River, above Lowell, is polluted by the sewage of Manchester, and of several smaller towns in New Hampshire; before reaching Lawrence, the sewage of Lowell, and a considerable amount of filth brought down by the Concord River, have been added. Haverhill, still lower down, gets the benefit also of the sewage of Lawrence, together with the refuse discharged into the Merrimac by the Shawshine and Spicket rivers, the latter of which, in summer, is rendered quite offensive by the filth from manufactories.[75]

[71] George E. Evans, "Engineer's Report," in *CDL*, 1872–3 (Lowell, 1873), 29–30.
[72] Water Board of the City of Lowell, *Third Annual Report* (1876), in *CDL*, 1875–6 (Lowell, 1876), 6–9.
[73] "Pure Water," Letter to the Editor, *Lowell Evening Citizen*, 7 Jan. 1891.
[74] Rice, *Report on a General System of Sewerage*, 6.
[75] Winsor, "Water-Supply, Drainage and Sewerage," 229.

As the report explained, this was already the case before the major cities on the Merrimack had completed comprehensive sewage systems. Rice was led astray by his faith in the capacity of rivers to purify themselves under all conditions. Later in the nineteenth century, urban wastewater further contaminated the water supplies of Lowell and Lawrence. This development caused a rise in waterborne disease, which, in turn, called the prevailing notion of river self-purification into question.

By the 1870s, the waters of the Merrimack valley were more thoroughly controlled and rationalized than ever before. Beneath the valley cities, water careened through pipes and tunnels, flowing off in a thousand different directions to help meet the needs of daily existence. The control of water on such a scale drew the valley together into a system of ecological interdependence. Common resources such as rivers, it is true, are by their very nature ecologically interdependent. This is an objective fact. But the perception people have of this interdependence can vary considerably.[76] By the latter part of the nineteenth century, New Englanders were becoming more aware of the way rivers tied together the fate of the people living in a watershed. In 1875, the MSBH sent questionnaires to the state's cities and towns asking if they knew "of any case where the sewage of one town pollutes the air or water of another." Fifteen towns, out of a total of 188, responded positively to the question.[77] Rising awareness of the interdependence of watersheds is also evident in the efforts to pass legislation regulating stream pollution.

The MSBH was the driving force behind water pollution legislation during the 1870s. Established in 1869 and headed by Henry I. Bowditch, a prominent New England physician and noted abolitionist, the MSBH was organized to investigate the sources of disease in the commonwealth. Early on, the board investigated conditions in slaughterhouses, and later studied the relationship between mortality and tenement life among the poor.[78] The seven-member board, made up of physicians, a

---

[76] My thoughts here derive from Thomas Haskell's discussion of the concept of interdependence. See Thomas L. Haskell, *The Emergence of Professional Social Science: The American Social Science Association and the Nineteenth-Century Crisis of Authority* (Urbana, Ill., 1977), ch. 2.

[77] Winsor, "Water-Supply, Drainage and Sewerage," 177, 228.

[78] Barbara Gutmann Rosenkrantz, *Public Health and the State: Changing Views in Massachusetts, 1842–1936* (Cambridge, Mass., 1972), 54–65.

mechanical engineer, and the well-known historian and journalist Richard Frothingham, began early in the 1870s to turn its
attention to water pollution.[79] In 1872, the Massachusetts legislature empowered the MSBH to survey the pollution of the
lower Miller's River in Cambridge and Somerville.[80] Three
years later, the board was authorized to examine the impact of
water pollution on the public health of the state.[81]

In the 1870s, after surveying some of the state's most industrial river basins, the MSBH began considering legislation to
prevent pollution.[82] With cities turning to rivers and lakes for
their public water supplies, the MSBH became alarmed over
the threat of pollution. In its report on the Nashua River basin
(see pp. 209–11), the board discussed recent water pollution
legislation in England.[83] In 1876, Britain passed an act
intended to curb river pollution. That legislation prohibited the
discharge into rivers of solid and liquid pollutants, including
sewage. As the MSBH noted, the act seemed a compromise
measure designed to meet the needs of a nation "not ready for
any more advanced measures."[84] The law had a number of problems. Polluters could continue to dump waste into streams provided they employed the best available means for purification.
Enforcement of the bill was left to inspectors who were not to
compromise the legitimate needs and interests of industry. And
the act provided no precise chemical definition of what constituted pollution. That omission left such matters for judges
to decide on an ad hoc basis.[85]

The rivers of Massachusetts, the board felt, were not nearly
as polluted as those in Britain. Yet there seemed a need for
legislation to prevent further contamination. Beginning in
1877, the board sent out surveys to 363 manufacturers located
in the state's principal cities and along its most industrial rivers.

---

[79] In the mid-1870s, the MSBH included Bowditch, Frothingham, John C. Hoadley, a
mechanical engineer, and Charles F. Folsom and Robert T. Davis, both of whom
were physicians, David L. Webster, and Thomas B. Newhall.

[80] Act of 6 May 1872, ch. 353, [1872] Mass. Acts & Resolves 320.

[81] Act of 8 May 1875, ch. 192, [1875] Mass. Acts & Resolves 785.

[82] Resolve of 27 Apr. 1877, ch. 56, [1877] Mass. Acts & Resolves 686. The word
"pollution" had long had moral and religious connotations stemming from its association with sin; by the latter part of the nineteenth century, the word took on its
modern physical meaning. See the *Oxford English Dictionary.*

[83] See Folsom, "Pollution of Streams," 73–9, for a discussion of the legislation and a
copy of the British statute.

[84] Ibid., 73.

[85] Hamlin, "What Becomes of Pollution?" 49–54.

The questionnaire asked if the manufacturer was troubled by pollution, what type of industry generated the worst pollution, and what they thought should be done to prevent it. The results were predictable: Two-thirds of the inquiries went unanswered. Of those, the board concluded, "it is probably safe to suppose that they are not enough troubled to enter any complaint." The survey was filled out by 119 manufacturers. Complaints were received from 14 mills – 5 of which were in the Nashua River valley – bothered by sewage and waste from textile, paper, and other factories.[86] Some respondents (38) thought a remedy to the problem of pollution was possible. (The report, however, does not mention their suggestions.) The rest either failed to answer this question, were unprepared to venture an opinion, or believed "nothing is better . . . than allowing the filth to run off by the streams."[87]

Heeding the interests of manufacturers, the board ruled out sweeping legislation severely restricting industrial pollution. "It would be unwise," wrote the board, "to place too great restrictions upon manufacturers by setting up, for all cases, some arbitrary standard of purity, which must be always followed, but which could not be enforced." Instead, the MSBH proposed "to regulate, rather than wholly prohibit, the contamination by filth of our waters."[88]

In 1878, "An Act Relative to the Pollution of Rivers, Streams and Ponds Used as Sources of Water Supply" passed into law.[89] The act was every bit as impotent as the board's comments would lead one to expect. The law prohibited the discharge of human waste, refuse, or other pollutants into any water within twenty miles of a public water supply. But one section of the law undermined its effect by failing to make the act retroactive. Those corporations or persons who had a right to pollute, either by prescription (long-continued use) or legislative grant, could continue to do so. The act merely prevented any further discharge of waste into water used for domestic purposes by those without any previous right to pollute. In all, the future of the state's water resources rested on a rather flimsy piece of legislation.

The law, in any case, would have no effect on the Merrimack

86 MSBH, *Ninth Annual Report* (1878), xv, 55–9.
87 Ibid., 66.
88 Ibid., 66–7.
89 Act of 26 Apr. 1878, ch. 183, [1878] Mass. Acts & Resolves 133.

River. That river, the Connecticut River, and part of the Concord River near Lowell were exempt from the legislation. Why these rivers were excluded is not entirely clear.[90] But a clue to the exemption of the Merrimack is found in statements the MSBH made when it proposed a draft of the law. In the board's view, it would be difficult for the huge cities on the lower Merrimack to meet any absolute standards for stopping pollution. A stringent law would force the city of Lowell, for example, to refrain from adding any additional sewers or water closets unless it could purify the resulting sewage. With as yet little firm understanding of how to purify sewage, the board implied that this would be too restrictive a measure. Still, the MSBH recommended no exemptions for specific rivers when it drafted the legislation.[91]

Hiram Mills, a member of the MSBH, later proposed a different view of the Merrimack River's exemption. According to Mills, it was pointless to stop pollution along the lower Merrimack when no regulations existed in New Hampshire to prevent its cities from sending waste into the river. New Hampshire did not even organize a state board of health until 1881. Moreover, since the large New Hampshire cities of Concord and Manchester drew their water from neighboring ponds, not from the Merrimack River, pollution was less of a concern than in Massachusetts. With no way of controlling pollution upstream in another state, the commonwealth, Mills believed, passed a law making the Merrimack River "a free receptacle for sewage."[92] Yet such an explanation does not make clear why the Concord River near Lowell qualified for exclusion. Whatever the case, the textile cities along the lower Merrimack were offered special treatment under the legislation.

There were further problems with the law. Nothing in the act, it was stated, would free polluters from the strictures of prevailing legal doctrine, presumably as articulated by the state's courts.[93] Those people, cities, or corporations allowed by the act to continue polluting could do so if they abided by the

[90] Evidently, when the bill was debated in the Massachusetts Senate, one senator voiced an objection to excluding the Merrimack River. There seems to be no record describing the substance of the objection. See "The Senate Debates the Water Supply," *Boston Daily Globe*, 5 Apr. 1878.

[91] MSBH, *Ninth Annual Report* (1878), 71–6.

[92] Hiram F. Mills, "Typhoid Fever in its Relation to Water Supplies," in MSBH, *Twenty-Second Annual Report* (1891), 533.

[93] Act of 26 Apr. 1878, ch. 183, [1878] Mass. Acts & Resolves 133, § 3.

precedents handed down by the Massachusetts court system. This meant, in effect, that the problematic doctrine of reasonable use – the centerpiece of water law at this time – would guide those who dumped pollution into rivers.

By the 1860s, as discussed in Chapter 5, New England's courts were promulgating the reasonable-use principle. In *Merrifield* v. *Lombard* (1866), the Massachusetts SJC applied the doctrine in a dispute over pollution.[94] The plaintiff, owner of a steam-powered factory along a river in Worcester, sued for an injunction to stop the defendant from discharging pollutants into the river. Evidently, the plaintiff believed his steam boilers were corroded by the pollution the defendant sent downstream. Chief Justice Bigelow, who wrote the court's opinion, granted the injunction. As he put it, any diversion of water

> which defiles and corrupts it to such a degree as essentially to impair its purity and prevent the use of it for any of the reasonable and proper purposes to which running water is usually applied, such as irrigation, the propulsion of machinery, or consumption for domestic use, is an infringement of the right of other owners of land through which a watercourse runs, and creates a nuisance for which those thereby injured are entitled to a remedy.[95]

The doctrine of reasonable use gave judges the freedom to decide what was in the public's best interests, to weigh the costs and benefits of pollution. Indeed, pollution cases were often excellent opportunities for judges to clearly articulate the doctrine.[96] This comes as no surprise. The adverse effects of pollution brought into sharp relief the costs of industrial production. This made the need for a balancing test, for a way to decide on the relative economic merits of river use – the basis of the reasonable-use rule – all the more compelling.

The reasonable-use principle, with its affinity for maximizing productive potential, could hardly be depended on to limit river pollution. Nor was the pollution act passed in 1878 effective in rescuing the state's waters from ecological degradation. The law did not prevent those with the right to pollute from continuing to do so; it excluded from its provisions three major

---

[94] *Merrifield* v. *Lombard*, 13 Allen at 16 (Mass. 1866); also see *Dwight Printing Co.* v. *City of Boston*, 122 Mass. 583, 589 (1877).

[95] *Merrifield* v. *Lombard*, 13 Allen at 17.

[96] See, e.g., *Hayes* v. *Waldron*, 44 N.H. 580 (1863), and *Snow* v. *Parsons*, 28 Vt. 459 (1856). *Hayes* v. *Waldron* (see Chapter 5), which clearly articulates the reasonable-use doctrine, involved the discharge of sawdust and wood shavings into a river. *Snow* v. *Parsons* dealt with pollution from a tannery.

industrial rivers; and it left decisions on the right to pollute to judges who for years had been propounding the problematic reasonable-use concept. Moreover, the law established no explicit penalties for those who polluted.[97] This emasculated piece of legislation gave state health officials little authority to regulate pollution. In the period following the passage of the act, writes the historian Barbara Rosenkrantz, "exhortations by the Board [MSBH] for cooperation from local boards of health and requests that manufacturers refrain from polluting water were frequently ignored or contested."[98]

In 1886, investigators for the MSBH reexamined the Merrimack's water at Lowell and Lawrence. Tests conducted in 1873 had found the water in acceptable drinking condition. Now increased contamination compromised the purity of water in these two cities and at towns further downstream, posing a threat to public health.[99] Mortality rates in Lowell from typhoid fever, a water-borne disease, peaked in 1870 and appear to have declined sharply over the decade as the city piped in water from the Merrimack (Table 7.6). But in the 1880s, contaminated water caused typhoid mortality rates to increase again; the rates reached epidemic proportions by 1890.

Typhoid fever is a bacterial disease. Its symptoms include diarrhea, abdominal pain, enlargement of the spleen, and prolonged fever. Nicknamed "the dirty hands disease," typhoid is spread through direct contact with the excrement of sufferers. It is also transmitted through linen, food, and frequently drinking water.[100] During the latter part of the nineteenth century, before adequate water and sewage treatment facilities existed, typhoid fever troubled most large American cities. In 1880, the cities of Atlanta, Louisville, Nashville, Newark, Philadelphia, Pittsburgh, Richmond, and Washington all had mortality rates of over 50 deaths per 100,000 from the disease. The disease emerged with particular virulence in Nashville and Pittsburgh, both with death rates over 130 for the year.[101]

---

[97] See Rosenkrantz, *Public Health and the State*, 82–3, on the difficulty of interpreting the law.

[98] Ibid., 87.

[99] MSBH, *Eighteenth Annual Report* (1887), xxxvii–xxxviii.

[100] Jean-Pierre Goubert, *The Conquest of Water: The Advent of Health in the Industrial Age*, trans. Andrew Wilson (Princeton, N.J., 1986), 48.

[101] See the typhoid fever mortality rates in Joel A. Tarr et al., "Water and Wastes: A Retrospective Assessment of Wastewater Technology in the United States, 1800–1932," *Technology and Culture* 25 (1984): 240.

Table 7.6. *Typhoid fever mortality: Lowell, Massachusetts,
1860–80*

Year	Mortality (per 100,000)
1860	29.87
1865	54.86
1870	80.63
1875	68.43
1880	38.67

Sources: Derived from Mass. Bureau of Statistics of Labor, *Census of the Commonwealth of Massachusetts 1905*, 4 vols. (Boston, 1908–10), 849–50; "Annual Report of the City Physician and Superintendent of Burials for the Year 1868," in *CDL*, 1868–9 (Lowell, 1869), 7; "Annual Report of the City Physician and Superintendent of Burials for the Year 1875," in *CDL*, 1875–6 (Lowell, 1876), 7; "Annual Report of the City Physician and Superintendent of Burials of the City of Lowell for the Year 1880," in *CDL*, 1880–1 (Lowell, 1881), 5.

An epidemic of typhoid fever struck the cities of Lowell and Lawrence in the winter of 1890–1. Between November and February, death rates from the disease in these cities ranged from 15.71 to 42.64. Mortality peaked in November in Lowell, and in the following two months downstream in Lawrence. There were 171 cases of the disease reported in Lowell for November alone.[102] According to Dr. H. P. Wolcott, chairman of the MSBH, this was "the most serious epidemic of typhoid fever within the commonwealth for many years."[103] Anxiety over the spread of the disease continued into the winter as newspaper reports speculated on the cause of the epidemic. Although some accounts suspected the city water supply, others blamed the putrefaction of human waste in the soil, which supposedly spread the disease through the air.[104] The latter theory was favored by most physicians during the second half of the nineteenth century. They believed diseases such as typhoid fever were transmitted through filth, through the stinking miasmas created as feces and other wastes putrefied.[105]

Some people, however, wondered if the Merrimack River –

---

[102] George C. Whipple, *Typhoid Fever: Its Causation, Transmission and Prevention* (New York, 1908), 150, 152.
[103] "The Water Question," *LET*, 13 Jan. 1891.
[104] "The Prevalence of Typhoid Fever," *Lowell Evening Mail*, 19 Dec. 1890; "The City Water," *LET*, 12 Jan. 1891.
[105] Tarr et al., "Water and Wastes," 232. Also see MSBH, "Inquiry into the Causes of Typhoid Fever," 110–79.

the primary source of water for both cities – was responsible for the epidemic. As one Lawrence observer wrote:

> Ten years ago the water of the Merrimac river was much purer than it is today. Every year sees more refuse in the river from cities above us, and the water is, of course, polluted all the more. . . . Lowell is only nine miles above us, and all the sewage from her 80,000 people finds an outlet into the Merrimac. In addition, too, the poisonous dyes and mill refuse also find an outlet in the river. This disease-laden water flows only nine miles, and then the people of Lawrence drink it. No wonder that disease and death follow.[106]

Newspapers in both cities demanded to know whether the Merrimack's water was safe to drink. After one Lowell paper carried a story blaming the Merrimack for the epidemic, the clerk of the water board wrote in to defend the purity of the city's water.[107] The *Lawrence Evening Tribune* questioned the competence of the MSBH, suggesting that the board had concealed information about the contamination of Lawrence's water supply.[108]

Since the 1878 water pollution act excluded the Merrimack River, the MSBH had no legal authority to regulate the water supplies at Lowell and Lawrence. But the life-threatening nature of the epidemic led to calls for state intervention. The *Tribune* demanded "either that the law prohibiting pollution of rivers and other sources of water supply shall be amended to include the Merrimack, or else that these cities and all others on the excepted rivers hereafter shall seek some other supply."[109]

A number of possible solutions were put forth to remedy the situation. One suggestion would have required cities to purify their sewage before disposing of it in the Merrimack. Questions were raised, however, about whether filtration would remove all the disease-causing impurities.[110] Under another plan, the city of Lawrence seriously considered abandoning the Merrimack and turning to two small bodies of water near the city.

---

106 "Fever Epidemic," *Lowell Daily News*, 26 Dec. 1890.
107 "Lowell's Drinking Water," Letter to the Editor, *Lowell Evening Mail*, 27 Dec. 1890. The previous week, the *Lowell Evening Citizen* ran an article anonymously quoting local physicians who attested to the gravity of the epidemic and implicated the city water. Unfortunately, copies of the paper from this time period have not survived. However, the *Lowell Evening Mail*, which accused the *Citizen* of sensational reporting, carries references to the article printed in the *Citizen*. See "Typhoid Fever and its Prevalence," and "The Citizen and the Typhoid Fever Scare," *Lowell Evening Mail*, 22, 26 Dec. 1890.
108 "Tribune Topics," *LET*, 24 Jan. 1891.
109 "The Water Question," *LET*, 13 Jan. 1891.
110 "Tribune Topics," *LET*, 26 Jan. 1891.

But the drainage areas of these ponds were far too meager and the supply insufficient to meet the needs of the large city.[111] Finally, one rather grandiose scheme called for an enormous sewer spanning from Concord, New Hampshire, "to the sea, to take the refuse matter from the many cities on the Merrimack and carry it to the ocean, where the chances of its ever becoming a menace to the public health would be exceedingly small."[112] In effect, the waters of the Merrimack River already acted as the enormous sewer envisioned. As one newspaper account concluded, the Merrimack River "as a source of domestic water supply . . . has lost its virtue because of the fact that to all intents and purposes it has become a gigantic sewer."[113]

With the health of thousands at stake, the MSBH finally intervened. By the latter part of the 1880s, the MSBH was better prepared to handle the threat pollution posed to public health. Legislation passed in 1886 gave the board authority for "the general oversight and care of all inland waters."[114] The new law also allowed the board to experiment with the best method for purifying sewage. The following year the Lawrence Experimental Station opened under the leadership of Hiram F. Mills, an Essex Company engineer appointed to head the MSBH's committee on water supply and sewage. Mills would play a key role in stemming the epidemic and creating a safe drinking water supply in the lower Merrimack valley.[115]

Hiram Mills had a long and distinguished career as a civil engineer. Early on, he had worked as an assistant engineer to James Kirkwood on the Brooklyn Waterworks and under James Francis at Lowell. Later he designed a number of dams, becoming well known as a hydraulic engineering consultant.[116] With tremendous self-possession at the age of just thirty-one, Mills composed a statement to the government of Maine urging it to develop the state's waterpower.[117] Boldly titled *Natural Resources and Their Development,* Mills's 1867 statement considered

---

[111] "Tribune Topics," *LET,* 26 Jan., 2 Feb. 1891; "The Water Question," *LET,* 7 Feb. 1891.

[112] "Tribune Topics," *LET,* 21 Jan. 1891.

[113] Ibid., 26 Jan. 1891.

[114] Act of 9 June 1886, ch. 274, [1886] Mass. Acts & Resolves 230.

[115] Rosenkrantz, *Public Health and the State,* 99.

[116] American Society of Civil Engineers, *A Biographical Dictionary of American Civil Engineers* (New York, 1972), 91–2.

[117] Hiram F. Mills, *Natural Resources and Their Development: Memorial of Hiram F. Mills, Civil Engineer, to the Governor and Council of Maine* (Augusta, Me., 1867).

the water resources of the entire country. The Northeast, he concluded, was remarkably well endowed with waterpower that if wisely handled, would offer the opportunity for production and wealth. "Seeing this," he wrote, "we can no longer be indifferent spectators, while our resources are running to waste."[118] Not one to sit idly, Mills soon worked to develop the waterpower in Bangor, Maine.[119]

As Mills saw it, there was much to recommend what the Boston Associates had done to the waters of New England.[120] The Associates followed through on precisely what he urged to the government of Maine: the systematic development of water for large-scale industrial production. Mills himself played a part in the Boston Associates' ventures. For a time, he helped maintain the waterpower system at Lowell before becoming an Essex Company engineer in 1869.[121] But as Mills grew older, he began to deal with the darker side of the industrial cities along the lower Merrimack River.

In November 1890, when deaths from typhoid fever in Lowell exceeded the number for the entire city of Boston (which had over five times the population), the MSBH began investigations under Mills's direction.[122] Mills compiled typhoid mortality rates for all the cities in the state between 1878 and 1889. He discovered that Holyoke, Lawrence, Lowell, and Chicopee – in that order – led the list. The highest mortality rates were found in manufacturing cities, especially in Lowell and Lawrence, where he uncovered even more disturbing data. In the period from 1886 to 1889, these two cities had "fifty per cent more deaths by typhoid fever, for the same population, than any other cities in the State."[123] Owing to the exemption under the 1878 pollution act, Lowell and Lawrence, as Mills pointed out, were the only cities in Massachusetts to take their water from a river into which sewage entered less than twenty miles above. Yet chemical analysis showed no more impurity in

[118] Ibid., 22.
[119] Mills was born in Bangor. American Society of Civil Engineers, *Biographical Dictionary*, 91.
[120] He even reviewed for readers of his memorial the rise of the Waltham–Lowell system along the Charles River and the move north into the Merrimack River valley. Mills, *Natural Resources and Their Development*, 18–19.
[121] *The National Cyclopaedia of American Biography*, 63 vols. (New York, 1891–1948; reprint, Ann Arbor, 1967–71), 12:71.
[122] Mills, "Typhoid Fever," 538.
[123] Ibid., 531–2.

the water passing Lawrence than in half the state's water supply.[124] How then could the elevated death rates from typhoid
fever be explained?

According to Mills, disease germs, and not chemical impurities, were responsible for the outbreak of typhoid fever.
This was a breakthrough in understanding the etiology of this
disease. Mills traced the disease to germs that went undetected
by chemical analysis and passed unnoticed through sewage-
polluted water.[125] Prevailing medical opinion held filthy conditions in cities responsible for epidemic disease. Mills and others
at the Lawrence Experimental Station overturned this view by
focusing on the microorganisms, such as the typhoid bacillus
(*Salmonella typhi*), responsible for specific diseases.

Stunning confirmation of the transmission of typhoid fever
through polluted water came from work done by the noted
biologist William T. Sedgwick. After sixty-four people in Lowell
died from typhoid fever in the autumn of 1890, Sedgwick began investigating the epidemic.[126] It must have been a nasty
job, tediously documenting where the cases occurred and how
those stricken with the disease might have passed the germs
into water consumed by others. But Sedgwick's study bespeaks
a man with enormous energy. Examining the city's water supply, he narrowed down the possibilities. The disease spread
through either the city's public water supply or through canal
water (or both, of course), two sources that came directly from
the Merrimack River. Canal water was pumped to the top of
many of the Lowell mills and used in water closets and for
washing. It was not meant for drinking since it was thoroughly
polluted with sewage and other wastes. But since the water was
easily accessible, and often cooler than water brought in pails
from wells and allowed to stand, workers often drank it.[127]

Sedgwick's inquiries led him to suspect the Merrimack River
as the route of transmission for the disease. With this in mind,
he tried to uncover how the epidemic began. Informed of an

[124] Ibid., 533.
[125] Ibid.
[126] See William T. Sedgwick, "On Recent Epidemics of Typhoid Fever in the Cities of
Lowell and Lawrence Due to Infected Water Supply; With Observations on Typhoid Fever in Other Cities and Towns of the Merrimack Valley, Especially Newburyport," in MSBH, *Twenty-Fourth Annual Report* (1893), 667–704. The figure for
deaths in Lowell from October through December 1890 is from Whipple, *Typhoid
Fever*, 152.
[127] Sedgwick, "Recent Epidemics of Typhoid Fever," 675–6, 678.

outbreak of typhoid fever in the village of North Chelmsford on the edge of the city of Lowell, Sedgwick investigated the case histories of individual patients. He ultimately traced the start of the epidemic to a single person, an infected ironworker who used a privy that emptied into Stoney Brook.[128] The now contaminated brook flowed into the Merrimack above where the city took its water supply.

In his report, Sedgwick went on to examine typhoid fever mortality rates for the Merrimack valley's main cities. He discovered that Lowell and Lawrence, with a combined population in 1890 about equal to the total for the cities of Concord, Nashua, Manchester, Haverhill, and Newburyport, had three times the rate of death from the disease between 1888 and 1893.[129] The reason for the elevated rates in the lower Merrimack cities, according to Sedgwick, stemmed from their dependence on sewage-polluted water – a dependence sanctioned by the legislation passed in 1878.[130] Lowell and Lawrence had high typhoid mortality rates, Sedgwick pointed out, because they have "constantly distributed to their citizens drinking water, unpurified, drawn from a stream originally pure but now grossly polluted with the crude sewage of several large cities and towns."[131]

To remedy this problem, Hiram Mills worked to purify the water of the lower Merrimack. Mills and his colleagues at the Lawrence Experimental Station discovered that filtering water through sand encouraged the biological decomposition of bacteria. Filtering the water in this way, according to Mills's observations, disinfected drinking water and led to a decline in mortality from typhoid fever.[132] The adoption of sand filtration eventually restored the purity of the Lawrence water supply. Moreover, by the 1890s, the Lawrence station was consulted by public health officials across the country for advice on purifying sewage-polluted water.[133] Evidently, Mills's career as a civil engineer had moved in a somewhat different direction, away from his earlier concern with using water for production and

---

[128] Ibid., 680–1.
[129] Ibid., 698.
[130] Ibid., 695.
[131] Ibid., 699.
[132] Hiram F. Mills, "The Filter of the Water Supply of the City of Lawrence and its Results," in MSBH, *Twenty-Fifth Annual Report* (1894), 560; also see Rosenkrantz, *Public Health and the State*, 101–2.
[133] Rosenkrantz, *Public Health and the State*, 101.

more toward purifying water for domestic use. Similarly, the lower Merrimack valley, long the center for industrial research on water, became a hub for experiments into the relationship between water and public health. This latter shift betokened the eclipse of the Waltham–Lowell system.

It is hardly coincidence that accounts for Hiram Mills's changing pursuits as a civil engineer. Indeed, his efforts on behalf of the Waltham–Lowell mills and his later focus on typhoid fever and drinking water were both based on the resolute control of New England's waterscape. Conquering water, whether for production or public health, was a driving ambition of an industrial culture desperate for progress.[134]

What Sedgwick, Mills, and others did at the Lawrence Experimental Station explained how contaminated water spread typhoid fever. Their work contributed toward the understanding of the etiology of this disease. It signified the triumph of science, the conquest of a threatening disease environment. But the studies conducted in the lower Merrimack valley were significant in other respects. They highlighted, more powerfully than ever before, the extraordinary interdependence of the watershed. An outbreak of typhoid fever in Lowell in 1892, Sedgwick soon discovered, led to epidemics downstream in Lawrence and Newburyport. This set of circumstances, as Sedgwick put it, "must be accounted one of the most interesting phenomena in our whole series of investigations, and may serve to confirm the truth of the saying that 'no river is long enough to purify itself.'"[135]

The idea that rivers purified themselves under any and all conditions, a notion often invoked earlier in the century, was now rejected. Flowing water, it was learned, offered no guaranteed immunity from the typhoid bacillus.[136] Indeed, just the opposite seemed the case. Rivers, in all their interdependence, tied the fate of the valley's population together. What takes place in a watershed seldom happens in isolation. This realization, struggling to become apparent for some time, now seemed much more obvious. Gone were the days when dumping waste into streams was the last one heard of the trouble.

[134] For a brilliant discussion of the mastery of water for health-related reasons, see Goubert, *Conquest of Water*.
[135] Sedgwick, "Recent Epidemics of Typhoid Fever," 704.
[136] See Mills, "Filter of the Water Supply," 560.

Part III

DECLINE

8

THE PRODUCTIVE VALUE
OF WATER

In the 1880s, George F. Swain called the Merrimack River "the most noted water-power stream of the world."[1] That remark was no mere hyperbole, coming as it did from a man who would soon become one of America's foremost civil engineers. A graduate of the Massachusetts Institute of Technology, Swain examined New England's waters as part of a U.S. government survey of the nation's waterpower. His object was to describe the region's rivers with an eye to their potential for future industrial development. But the plentiful waters of the Merrimack River, he discovered, had already been wisely employed. The Merrimack, he wrote, probably had "more power utilized than in any other drainage basin of equal size in America."[2] As Swain realized, the Merrimack valley had long been the heartland of waterpowered industry. But its years of success could not and did not last forever.

By the time Swain made his report, economic transformation had left a deep imprint on the Merrimack valley. Swain noted the physical features of the watershed, a landscape now largely devoid of forest. Indeed, except for the upper reaches of the watershed that were nestled in mountainous terrain, the basin was no longer thick with woodland. The forest had been cut down, in part to make room for the valley's now populous urban centers. There were other signs of how progress had reshaped the valley. Upstream at the New Hampshire lakes, the Lake Company had been fortifying the Merrimack valley's ca-

---

[1] George F. Swain, "Water-Power of Eastern New England," in U.S. Department of Interior, Census Office, *Reports on the Water-Power of the United States*, vols. 16 and 17 (Washington, D.C., 1885–7), 16:23 (hereafter cited as Swain, "Water-Power of Eastern New England").

[2] Ibid., 23. Swain computed the amount of waterpower utilized per square mile of drainage basin for some of the larger streams in the nation. The Merrimack River ranked third behind the Blackstone and Thames rivers. See ibid., xxxix.

pacity to store water. Swain mentioned that other streams offered an ample or better capacity for storing water, but "there are probably none in which those advantages have been so extensively and systematically improved."[3]

Swain saw a waterscape finely tuned to factory production: water controlled, rationalized, and distributed to meet the needs of manufacturing. As the water coursed downstream, it passed through the valley's centers of hydraulic innovation, the sprawling cities of Manchester, Lowell, and Lawrence. According to Swain's figures, the three factory towns met roughly two-thirds of their combined energy needs with water. The balance of their energy came from an alternative energy source: steam power.[4]

The roots of the Waltham–Lowell system's eventual obsolescence dated from the 1830s. With the opening of the coal fields of eastern Pennsylvania, anthracite became widely available to industry. The new fuel helped transform metal making in America, providing manufacturers with a reliable supply of iron and expanding the nation's industrial base. Coal, moreover, could also be used to generate steam power, freeing American industry from the need to be near available waterpower. This changed the contours of industrial geography, as manufacturers found it more profitable to build steam-powered factories closer to markets and abundant supplies of labor. By the mid-nineteenth century, integrated steam-powered textile factories were constructed along the Atlantic coast from Connecticut to New Hampshire.[5] Long the center of industrial innovation, the outstanding factory towns of the Merrimack valley now had formidable rivals, among them the textile city of Fall River, Massachusetts.

Tucked away in the southeastern part of the state, Fall River offered a bold new challenge to Lowell's industrial hegemony.

---

[3] Ibid., 23–4.
[4] The following table is derived from ibid., 27, 33, 37:

	Mill powers owned	Gross HP of water	HP of steam
Lawrence, Mass.	128	10,910	4,200
Lowell, Mass.	139 11/30	11,845	13,940
Manchester, N.H.	184	15,890	1,450

[5] Alfred D. Chandler, Jr., *The Visible Hand: The Managerial Revolution in American Business* (Cambridge, Mass., 1977), 76–7.

Originally dependent on the water of the Quequechan River for its power, the town experienced spectacular growth as it converted rapidly to steam. The Fall River mills doubled their number of spindles between 1855 and 1865. By 1875, they led the nation with over 1.2 million spindles, almost twice the number operating at Lowell.[6]

The dominant force in textile production for decades, Lowell succumbed to the advancing trend of steam-powered factories. Its long-term use of the Merrimack River no longer conferred a comparative advantage. Steam power comprised 96 percent of the increase in horsepower available to American manufacturers in the period from 1869 to 1889.[7] Ultimately, even the Lowell mills shifted to steam. Between 1867 and 1880, the number of steam engines more than doubled there, rising from thirty-one to seventy-three. By the time of Swain's report, Lowell employed more steam power than waterpower, a trend that continued over the next ten years.[8] Although the mills at Manchester and Lawrence were slower to adopt steam than those at Lowell, it was clear that waterpowered factories were becoming obsolete.[9]

Still, the beleaguered Waltham–Lowell mills left their mark on the Merrimack valley. There were, of course, the enormous changes they had made to the waterscape, the thorough control of water and the ecological consequences that resulted. But their legacy also had another dimension. The system itself was waning, but its instrumental vision of nature survived intact.

The present chapter examines the conflict that developed over water as the Waltham–Lowell system declined. This conflict was shaped by the larger cultural context that defined how New Englanders behaved toward both one another and the natural order. Economic, legal, and social change had combined over the course of the century to effect a stunning cultural transformation, a shift in the way people understood their relationship with nature. This understanding of nature urged

[6] Thomas Russell Smith, *The Cotton Textile Industry of Fall River, Massachusetts: A Study of Industrial Localization* (New York, 1944), 48; Louis C. Hunter, *Waterpower in the Century of the Steam Engine,* A History of Industrial Power in the United States, 1780–1930, vol. 1 (Charlottesville, Va., 1979), 494.

[7] Hunter, *Waterpower,* 481.

[8] *Statistics of Lowell Manufactures, January 1867* (Lowell, 1867); *Statistics of Manufactures in Lowell and Neighboring Towns, January, 1880* (Lowell, 1880); *Annual Statistics of Manufactures in Lowell and Neighboring Towns, January 1890* (Lowell, 1890).

[9] Swain, "Water-Power of Eastern New England," 27, 37.

the control of water for economic gain. It viewed water in terms of its productive value – a perception of the natural world with powerful and sweeping appeal. There emerged one main, overriding concern: to use water to promote economic growth, to employ it productively for profit. It only remained to be seen who would benefit as water's economic potential was maximized.

The Lake Company had consolidated its hold over New Hampshire's water by the 1860s, surviving the threat posed by a tenacious opposition. Power once attained is not easily given up, and as the century wore on, the company sought to retain its control over water in the face of continued state and local resistance. The Lake Company's inordinate thirst for and thorough mastery over water was a powerful reminder of the commanding presence the industrial interests at Lowell and Lawrence had in New England. To some at least, that presence had become far too burdensome.

Between 1866 and 1867, ten cases were brought against the Lake Company for flooding land on the shores of Lake Winnipesaukee – the start of a long, embittered legal battle.[10] In 1870, the town of Bristol, New Hampshire, claimed the company's water management at Newfound Lake had destroyed a public highway.[11] That same year, mill owners along Newfound River sued to recover damages caused by the company's handling of the water flowing to their mill. Before bringing legal action, the mill owners broke into the Lake Company's property and hoisted the gates on the dam.[12]

The litigation continued to mount. By 1872, the company was entangled in a remarkable twenty-six cases over water. In only two of those cases – both for trespass at the Newfound dam – was the Lake Company the plaintiff.[13] The rest found the company defending itself against the region's disgruntled – people who felt victimized by what the company had done to the water and to them. Three cases are worth singling out for consideration. Two were test cases upon which eighteen of the other suits rested. The final case was a suit in equity (all the

---

[10] J. French to H. Bartlett, 9 Aug. 1871, item 204, EC Papers, MATH.
[11] Report of Arbitrators, Suit of the Town of Bristol, 7 Sept. 1872, item 204, EC Papers.
[12] See *Holden v. Lake Co.*, 53 N.H. 552–4 (1873).
[13] Suits, June 1872, item 204, EC Papers.

others were actions at law), which was important because it suggested just how strong-willed the Lake Company had become.

The first case, *Gilford* v. *Winnipiseogee Lake Co.*, came before the New Hampshire SJC in 1872.[14] The town of Gilford sued because of the flooding of its land above the Lake Village Dam; in defense, the company claimed a prescriptive right. Gilford won in the lower court based on the jury's answers to three questions posed by the court. The questions were designed to determine whether the company actually had a prescriptive right to flood the land. Since the town claimed damages beginning in 1860, the company had to show that it had flooded the land for at least twenty years (the period necessary to maintain a prescriptive claim) prior to that time. The jury was not persuaded by the company's case. In their view, the Lake Company, before its control by the Boston Associates in 1847, had intended to keep the water below the top of the dam (question one), had indeed done so (question two), and had not consistently raised the water high enough on the dam to give surrounding landowners sufficient notice that the company claimed the right to flood (question three). The jury concluded, in short, that the company had not established a prescriptive right to flood the land. They found in favor of the town and assessed $616 in damages against the company.[15]

The decision could hardly have surprised the Lake Company's agent, Josiah French. When trial began in the summer of 1871, French questioned whether an impartial jury could be found.[16] After the verdict, he wrote: "It is impossible for the Company to get a fair trial on any flowage case, as there are so many who have shore, or have relatives who have, that a jury cannot be got in any County where the Company have water rights, that will not be so prejudiced, as to go against the Company, right or wrong."[17]

The company's case fell on equally unsympathetic ears the following year, when the verdict was appealed. In its opinion, the SJC established two main points of law that doomed the company's cause. First, merely erecting a dam and maintaining it for twenty years, they held, could not by itself establish a

---

[14] *Gilford* v. *Winnipiseogee Lake Co.*, 52 N.H. 262 (1872).

[15] Ibid., 262–5; Charles Storrow Diary, 14 Sept. 1871, item 204, EC Papers.

[16] J. French to H. Bartlett, 9 Aug. 1871.

[17] Ibid., 1 Sept. 1871, item 204, EC Papers.

prescriptive right to flood land. In their words, "Until the defendants actually raised the water on the plaintiffs' land by
means of their dam, they did nothing for which the plaintiffs
could have recovered nominal damages, 'in the assertion and
vindication of an invaded right.' "[18] For the company to have
established a prescriptive claim, it needed to manage the water
in a way that infringed on the rights of landowners. If the
landowners acquiesced by failing to take action against it for
twenty years, the prescriptive right to flood was established.

Second, the court ruled that a milldam owner did not have to
flood the land claimed every day for twenty years to gain a
prescriptive right. But it was necessary that the nature and
frequency of the use "give notice to the land-owner that the
right is being claimed against him."[19] On each of these
grounds, the Lake Company failed to adequately establish its
prescriptive right.

From a legal perspective, there was nothing terribly radical
about the opinion. It fell neatly within the trend, evident in
New Hampshire law since 1830, to make adverse use the sole
criterion for establishing a prescriptive claim. The legal rule of
adverse use had benefited the Lake Company in *Lake Company*
v. *Young* (see Chapter 5), but in this case, the doctrine was far
less congenial to its interests. The court in *Gilford* v. *Lake Co.* applied a stricter standard for demonstrating a prescriptive claim
to water. It agreed with *Lake Company* v. *Young*, which stated that
the use did not have to be continuous, yet it added a more
exacting limitation. Evidence of a prescriptive right could in
part be deduced by asking a simple question: "Has the landowner had reasonable cause, from the actual use made of his
land, to believe, during the entire period of twenty years, that
the right was being claimed against him?"[20] Despite the company's best efforts to finesse the issue, it clearly held no prescriptive right to flood under this more stringent limitation.
Losing the case made the company vulnerable in eleven other
cases that hinged on the outcome. The company eventually
made cash payments to settle those lawsuits.[21]

The Lake Company's management of the water at Newfound
Lake was at issue in the second test case, *Holden* v. *Lake Company*

[18] *Gilford* v. *Lake Co.*, 52 N.H. at 265.
[19] Ibid., 266.
[20] Ibid., 267.
[21] "Memo for Annual Meeting of June 1874," item 204, EC Papers.

(1873). Benjamin Holden and the others who joined him in the suit owned a woolen mill below the company's dam at the outlet of the lake. They claimed the company failed to provide them with adequate water in the summer to properly carry on their business. Before reaching the SJC, the case was heard by a jury that ruled on whether the Lake Company had made "reasonable use" of the water. The judge gave the jury an example, for purposes of comparison, of what constituted unreasonable water use:

> Suppose A becomes the owner of the outlet of a great lake, from which flows a large river, furnishing the power to drive many mills that have been erected on it by owners below; that the dam is of such a character and construction that the owner of it can control the flow of water in the stream; that he shuts down his gates and keeps them closed twelve out of every twenty-four hours, thereby entirely stopping the flow of water in the stream below, and during the other twelve hours raises them so as to allow the accumulated waters of twenty-four hours to escape in twelve, thereby causing an irregular and intermittent flow, so that the water comes down to the proprietors below in instalments [sic] or gusts of twelve hours each: I should suppose, in a case of that kind, the jury might not find it a reasonable use.[22]

Of course, the Lake Company had done precisely what the judge described. The need for a uniform flow compelled the company to manage the water in just that way. The procedure served the best interests of the mills below the dam only if their production schedules matched the company's water management. The tendentious jury instructions virtually guaranteed a victory for the plaintiffs. When the company lost the case, Charles Storrow wrote: "Chief argument against the Co. being that they are rich & robbers – came to N.H. to steal the property of the natives & usurp rights. Verdict against us for 2615.75 – perfectly injust [sic] & iniquitous."[23]

In 1873, the case came before the SJC. The Lake Company's lawyers were outraged by the instructions. In their view "the presiding justice proceeded to embody the defendants' precise case in the form of an illustration . . . and then gravely told the jury he should suppose, in such a case, they might not find the use a reasonable one! In other words, the judge had nothing to do with the reasonableness of the defendants' use, yet he

22 *Holden v. Lake Co.*, 53 N.H. at 555.
23 Charles Storrow Diary, 18, 19, 20 Nov. 1872.

thought it was unreasonable, and should suppose the jury would think so too!"[24] Apparently, the justices of the SJC did not see things this way. The jury instructions may have seemed prejudicial, but the court did not find fault with them. "If the case supposed was correct," wrote Chief Justice Sargent, "then the defendants' misfortune was, that the evidence against the defendants was so plain as to amount to simple demonstration."[25]

Fearing the prejudiced legal climate, Josiah French, Charles Storrow, and James Francis met in 1873 to contemplate divesting the Lake Company property. Desperation filled the air as the Lake Company's empire of water seemed about to crumble. The three men gathered, in Storrow's words, to consider the "general policy of the company under the vexatious opposition & injustice to which we are now exposed, under the recent action of Juries & Courts in N. Hampshire." They believed it wise to leave the state of New Hampshire, to sell off most of the Lake Company's property.[26] Not long before, such thoughts of retrenchment were buried beneath a much bolder attitude toward water and those who dared to oppose the company's plans. But by the 1870s, that attitude was fading fast before a vigorous legal challenge.

The third and final case – *Cole* v. *Lake Company* – capped the company's string of legal failures. Benjamin J. Cole, the plaintiff, was a manufacturer who had lived in Lake Village since age thirteen. His father, Isaac, opened a foundry in the town that had passed to Benjamin Cole and his brothers.[27] His brothers eventually left the foundry, which produced stoves and farming tools, and Benjamin Cole continued the business by himself. In 1848, he diversified his interests by serving as a founder and president of the Winnipiseogee Steamboat Company, builders of the first passenger steamboat on Lake Winnipesaukee (Figure 8.1). Over the years, Cole promoted the local economy of the village through his business pursuits, constructing over sixty buildings. One biographer described him as "public-spirited, benevolent, and mindful of the welfare of his numerous employés, and the citizens generally."[28] As he tended to business,

---

[24] *Holden* v. *Lake Co.*, 53 N.H. at 557.

[25] Ibid., 561.

[26] Charles Storrow Diary, 17 June 1873.

[27] Martin A. Haynes, *Historical Sketches of Lakeport New Hampshire Formerly Lake Village, Now the Sixth Ward of Laconia* (Lakeport, N.H., 1915), 12.

[28] Charles W. Vaughan, comp., *The Illustrated Laconian* (Laconia, N.H., 1899), 48.

Figure 8.1. *Lady of the Lake.* Stereograph, c. 1880. This steamboat, first built by Benjamin Cole and the Winnipiseogee Steamboat Company in 1848, is shown here after it was purchased by the Boston, Concord, & Montreal Railroad. (Courtesy of the New Hampshire Historical Society.)

however, Cole found himself increasingly at odds with the town's notorious corporate entity.

Under an agreement signed in 1852, Cole and his partners leased water from the Lake Company to run their iron foundry and machine shop. In the lease, the company reserved the right to control the water in the Winnipesaukee River and the lakes and bays above.[29] According to Cole, the company denied him access to the water by erecting a partition that reduced the water in his canal by as much as two feet. In an effort to get the company to honor the lease, Cole brought a lawsuit that reached the SJC during its 1874 equity term. The court ruled in his favor. According to the opinion, the company had no right to cut off the water in the canal leading to Cole's premises. The lease reserved for the company the right to control the water above the dam. But obstructing water in the canal, the

[29] See the facts in *Cole v. Lake Co.*, 54 N.H. 242, 244 (1874).

court concluded, was "in no way connected with the exercise of a general control of the water" above.[30]

For the moment, the decision in *Cole* v. *Lake Company* marked the end of the company's legal travail. By the time the case was settled in October 1874, the Lake Company had managed, largely by cash payments, to appease the twenty-odd plaintiffs who challenged it. And once it decided to drop the two cases it had brought for trespass, no litigation remained.[31] Why did the Lake Company fail so miserably against the legal onslaught?

Defeat stemmed, in all likelihood, from the sheer audacity of the company's water management. This was especially evident in Cole's case in which the company clearly violated the terms of the lease. The Lake Company, of course, had always been bold. From the start, the entire effort to control the New Hampshire lakes evinced a confident, stubborn impulse to subdue water and those who interfered with the company's management of it. But by the 1870s, it appeared that the company had overstepped the bounds of propriety and strayed too far from any reasonable standard of conduct. Nor is there reason to doubt the assessments of French and Storrow: The Lake Company now confronted a court system made up of juries and judges – especially if the *Holden* case is any indication – more inclined than ever to rule against it. Whatever the reason for the backlash, it seemed certain to continue so long as the Lake Company held a tight grip over the region's substantial reserves of water.

Eighteen years after the flagrant attack on the Lake Village Dam, it happened again. Early on Sunday, 20 May 1877, a crowd met at the dam and began forcing off planks to make

---

[30] Ibid., 274. The Lake Co. used two lines of argument in its defense. The court rejected its first argument that the lease permitted it to withhold all the water at its pleasure. But the company also claimed that the lease was a tenancy at will (where premises are held for no fixed term and can be terminated by the lessor when he or she wishes). The court ruled, however, that the agreement was not a tenancy at will but a fee simple – a perpetual right, in this case, to lease the property. The ruling here was controversial. To secure a fee simple under the common law, certain language was necessary, including the explicit use of the word "heirs" to denote the perpetual nature of the agreement. The contract signed by Cole and the company, however, did not use such language, forcing the court to declare that the need for the word "heirs" was irrelevant under the present circumstances. In the court's opinion, the need for such language "was nothing more nor less than the practice of the feudal sovereign. . . . Its origin, purpose, and history show it to be in no way adapted to our institutions, system of government, or condition of society." Ibid., 285.

[31] "Memo for Annual Meeting of June 1874."

Figure 8.2. The Lake Village Dam. Stereograph, c. 1880. (Courtesy of the Museum of American Textile History.)

way for a run of logs. Josiah French had since retired from the company and died the year before. The new agent, Jotham P. Hutchinson, went to the dam to order the intruders away.[32] But the crowd persisted, and despite the efforts of Hutchinson and his men, the dam's attackers opened a passage for the logs (Figure 8.2). Hutchinson later claimed to be completely surprised by the incident, although he recalled someone – though neither Gardner Cook or John Aldrich, who owned the logs – seeking permission to let logs through the dam. He refused the request and put it out of his mind.[33]

Hutchinson turned to Ellery A. Hibbard to assess the legality

[32] Hutchinson was a lawyer and the brother-in-law of Thomas J. Whipple, the Lake Co.'s attorney in *Lake Company v. Young*.
[33] C. Storrow, Notes on Cook & Aldrich, 1877, item 204, EC Papers.

of the attack.[34] The main question of law centered on whether the Winnipesaukee River between Paugus Bay (Long Bay) and Opechee Bay (Round Bay), where the dam was located, could be considered a public highway. If it was, then Cook and Aldrich were legally entitled to abate the nuisance (i.e., the dam) to get their logs through.[35] Under the common law, all points in a river where the tide ebbed and flowed were considered navigable (water thus influenced by the ocean's tide was deemed a part of the fully navigable sea). But New Hampshire law had for some time recognized even non-navigable rivers as public highways if they had long been used for that purpose. In other words, a prescriptive right to use a river as a public highway could be acquired by long use.[36] A recent legal decision held further that adverse use alone was not the only way for a river to become part of the public domain. The case of *Thompson* v. *Androscoggin River Improvement Co.* (1874) decreed a river to be public if in its natural condition it could be used for transportation.[37] In Hibbard's view, Winnipesaukee River, "although not navigable in the common law sense is . . . a public highway for the transportation of lumber to market to the extent that it was capable of being used for the purpose in its natural state."[38]

It was possible, however, for this public right to be lost. This could happen, Hibbard explained, if dam owners "gained a right by twenty years of adverse uninterrupted occupation, to prevent its use for the purpose [of transporting logs]." According to Hutchinson, no logs had been allowed to pass over the dam for twenty-six years, more than enough time to extinguish the public right to use the river. But Hibbard pointed to a statute in effect since 1868 prohibiting a dam owner from gaining title through prescription to obstruct navigation.[39] Thus, the company needed to demonstrate that no logs were passed for twenty years before 1868. If no logs were passed for twenty-

---

[34] Hibbard was the lawyer for the defendants in *Lake Company* v. *Young*.

[35] E. A. Hibbard to J. P. Hutchinson, 18 June 1877, item 204, EC Papers.

[36] See, e.g., *Scott* v. *Willson*, 3 N.H. 321 (1825). For the common-law definition of a navigable river, see Joseph K. Angell, *A Treatise on the Right of Property in Tide Waters and in the Soil and Shores Thereof*, 2d ed. (Boston, 1847), 75–6.

[37] *Thompson* v. *Androscoggin River Improvement Co.*, 54 N.H. 545, 548–9 (1874). Justice Charles Doe's opinion held further that parties need not prove that the waterway in its natural condition made it a public highway. A judge, he ruled, can take judicial notice (where a judge without the production of evidence decides to accept certain truths as given) based on his own knowledge and experience.

[38] E. A. Hibbard to J. P. Hutchinson, 18 June 1877.

[39] Ibid.

six years, as Hutchinson held, the company still fell three years short of the twenty-year minimum.[40] From a legal standpoint, the company stood on shaky ground. Hibbard ventured no opinion on whether the company should try to pursue its interests in court or simply put in a passage to accommodate the loggers.[41]

Ultimately, the Lake Company bowed to prudence and constructed a sluiceway in the fall of 1877.[42] The entire incident, nevertheless, completely enraged Charles Storrow. He preferred a more genteel world, one that chose to settle differences in words and writing, not by violence. If the Lake Company had indeed acted unlawfully by refusing the passage of logs, Storrow believed ample legal recourse existed. Nothing the company did or failed to do would "justify violence on the Lord's day – without notice – and the forcible taking possession of the company's property, and its injury in this manner."[43]

If Charles Storrow preferred people who calmly aired their grievances, he would soon find plenty of them. In June 1877, 1,180 citizens of Belknap and Carroll counties petitioned the New Hampshire legislature.[44] The petitioners included "many of the most respectable and influential citizens" in the region affected by the Lake Company's water management.[45] They believed the Lake Company had "grown to be a large and powerful corporation, exercising unwarranted and arbitrary powers over a very considerable portion of the waters and territory of this State." Although the company had been chartered as a manufacturing enterprise, as far as the petitioners could tell, it had produced virtually nothing. On the contrary, "its management and affairs are so conducted as to greatly injure the business and prosperity of the citizens of this State." They asked the legislature to repeal the company's act of incorporation.[46]

In style and demand, the petition resembled the one signed

---

[40] C. Storrow, *Notes on Cook & Aldrich.*
[41] E. A. Hibbard to J. P. Hutchinson, 18 June 1877.
[42] *LVT,* 13 Oct. 1877.
[43] C. Storrow, *Notes on Cook & Aldrich.*
[44] *Charter of the Winnipiseogee Lake Company together with the Amendment made in 1846, and A Copy of the Petition presented to the Legislature in 1877* (Lake Village, N.H., 1878), 10–11 (hereafter cited as *Petition*).
[45] Committee Report of the N.H. House of Representatives on Petition in Regard to the Lake Company, 1877, vol. 46, file 275, PLC Papers, BL.
[46] *Petition,* 11.

by almost as many people exactly twenty years earlier (see p. 117). This time, however, the petitioners attracted far more attention from the legislature. Three New Hampshire lawyers were appointed to investigate the matter: Francis Faulkner, John Farr, and Charles W. Woodman. Over the next two years, they heard testimony from both sides as the history of the Lake Company came under close public scrutiny. For the company, the outcome would decide its fate in this state. But the proceeding would also render judgment on the past, on what the company had done to the waters of New Hampshire and whether it had acted in the best interests of the people who lived here.

On Wednesday, 20 February 1878, the legislative committee convened at Mount Belknap Hall in Lake Village. The three committee members and the lawyers representing the Lake Company, Ellery Hibbard and Thomas J. Whipple, gathered for the start of the session. The petitioners arranged to have the firm of Rand, Albin, & Clark speak on their behalf.[47] Edward D. Rand played the leading role in the petitioners' cause. Born in Bath, Maine, in 1821 and a graduate of Wesleyan University, Rand was a skilled orator, a man whose "sentences were well moulded and his words fitly chosen."[48]

At 9:45 in the morning, as local citizens filled the hall, Rand gave his opening statement. He began by reading the Lake Company's original act of incorporation (granted in 1831) and noted the additional measure passed in 1846 increasing the company's capital stock to $1 million. In his view, the company had no right to use the water of Lake Winnipesaukee for any other purpose than the manufacturing of textiles. It was sheer deception for the company to represent itself as a manufacturing concern. It was instead a reservoir company, designed to manage the water for purposes outside the state of New Hampshire. Of the Lake Company's four hundred shares of stock, he pointed out, half were owned by the textile companies in Lowell and half by the Essex Company in Lawrence. Beyond violating the terms of the charter, he continued, the company had also stifled the economic potential of Lake Village. Cautioning local citizens not to take offense, he believed Lake Village to be a rather sorry excuse for a business community. Under the conditions imposed by the Lake Company, it proved impossible for industry to thrive. He had never seen a lease for

[47] "The Lake Company Investigation," *LVT*, 23 Feb. 1878.
[48] Charles H. Bell, *The Bench and Bar of New Hampshire* (Boston, 1894), 604–5.

waterpower in which the company did not reserve for itself ultimate control over the water. He claimed the company also blackmailed its waterpower tenants by forcing them to pay a portion of the company's tax bill. Beyond the harm it had caused Lake Village, the company's water management also injured those owning meadows along the Winnipesaukee and Merrimack rivers. In addition, by retaining the water in Lake Winnipesaukee, the company jeopardized public health by hindering proper drainage. Finally, the company stood accused of endangering navigation on Lake Winnipesaukee.[49]

In all, this was quite a searing indictment. During the following weeks, Rand and his associates called witnesses to testify. The witnesses represented a cross section of the economic life of the region. There were manufacturers, loggers, steamboat pilots, mechanics, and farmers. Among them was R. M. Appleton, a manufacturer who rented waterpower from the Lake Company since 1870 and blamed it for the shortage of water at his mill. A yarn worker in the village named Lyman Pulcifer said that the company had not manufactured anything in thirty years except briefly in 1858.[50] According to Moses Gordon, a farmer who lived nearby, Josiah French had once admitted to him that he cared not for the welfare of Lake Village so long as he could do with the water what he wished. Winborn Sanborn, a steamboat captain on Lake Winnipesaukee, recalled complaining to the Lake Company about the low water in the lake that hindered navigation.[51] And the local physician, Oliver Gross, attributed the prevalence of typhoid fever in the village to the company's policy of drawing down water.[52]

John Aldrich and Gardner Cook testified on the latest incident at the Lake Village Dam. As Aldrich explained, they had bought lumber around the lake in 1876 and 1877 and expected to transport it to Lake Village. There, through an earlier agreement with Jotham Hutchinson, they planned to saw it at one of the Lake Company's sawmills. But when the lumber arrived, Hutchinson said he had orders not to lease the mill on a short-term basis and proposed instead to saw it for them. Because

---

[49] "The Lake Company Investigation," *LVT*, 23 Feb. 1878; "The Lake Company Investigation," *Laconia Democrat*, 23 Feb. 1878.
[50] "The Lake Company Investigation," *LVT*, 23 Feb. 1878.
[51] "The Lake Company, Continuation of the Hearing Before the Commissioners," *LVT*, 2 Mar. 1878 (hereafter cited as "Hearing," 2 Mar. 1878).
[52] "The Lake Company, Continuation of the Hearing Before the Commissioners," *LVT*, 9 Mar. 1878 (hereafter cited as "Hearing," 9 Mar. 1878).

Hutchinson insisted on having it all cut at one time, rather than to meet specific orders as Cook and Aldrich desired, they decided to run the logs through the dam on route to their own mill downstream in Laconia.[53] This version of events makes the breaching of the dam seem less an act of gratuitous violence – as Storrow would have it – than a decision of last resort. But since the Lake Company eventually agreed to have a sluiceway built, as Aldrich admitted, the company may have repaired its public image.

For their part, the Lake Company's lawyers sought to portray the company as an important contributor to the village's economy. They accomplished this by having witnesses juxtapose a quaint image of the town as a small, undeveloped New England community before the founding of the Lake Company, with its more recent growth in population and business. Above all, the company's lawyers focused on their client's positive role in boosting the economic potential of the region's water.

When Benjamin Cole testified, he recalled periods of substantial economic growth in the fifty years he had lived in Lake Village. But as far as he could tell, the Lake Company had done little to aid the local economy. It had not constructed any new buildings in the town, except for one store.[54] In regard to water, Cole leveled heavy criticism. Although Cole was sometimes short of water, French at one point accused him of drawing too much. When the company learned that Cole intended to take advantage of all the waterpower to which he was entitled, it hauled rocks into the canal to cut off his water supply. Indeed, to hear Cole recount how French once approached him for a rent increase – "'Mr. Cole, you are not paying enough rent.'" – is to suspect the rank manipulation involved here.[55] Yet in the end, Cole admitted under cross examination that the company had greatly improved the waterpower of the village.[56] It was a stunning admission from such a fervent critic of the company. No matter how exploitative the Lake Company had been, it was likely to be cleared of the most troubling of the petitioners' charges if it showed that it helped the overall economy.

Other people besides Cole offered conflicting testimony

---

[53] "Hearing," 2 Mar. 1878.
[54] "Hearing," 9 Mar. 1878.
[55] "The Lake Company Investigation," *Laconia Democrat*, 9 Mar. 1878.
[56] "Hearing," 9 Mar. 1878.

about the Lake Company. Indeed, the company seemed to evoke contradiction among those who came into contact with it, especially among proprietors who rented its waterpower. The experiences of Moses Sargent are noteworthy in this regard. When the Lake Company sued George Young for overdue rent, Sargent, who rented the woolen mill nearby, disputed Young's claim that the company denied him water. This time, Sargent testified against the company. However he felt about the Lake Company back then, he was now more critical of its behavior. His own experience as a tenant of the company had been burdensome. At one point, Josiah French sought to charge him additional rent to help subsidize the company's local tax bill. In Sargent's view, the exorbitant rent eventually forced him out of business.[57]

Sargent had clearly soured on the Lake Company. But there was a day, in the not too distant past, when he saw fit to help defend company property. According to Isaac Colby – a Lake Company mechanic who also testified – Sargent had helped protect the dam when George Young and others arrived to destroy it in 1859.[58] Perhaps at age seventy-four, Sargent's memory of that day had faded. It is hard to say how he reconciled his earlier defense of the dam with his present criticism of the company. Allegiance is often a fickle thing. But the revelation of these facts must have cast a shadow over Sargent's testimony and could not have persuaded the commissioners of the company's guilt. The company seems to have played an ambiguous role in the region – one that could lead to contradiction in the course of a single lifetime.

In March 1878, Charles Storrow prepared to defend the company before the commission. The company had portrayed itself at the hearings as vital to the region's economic progress. This was part of a more general strategy employed from the start of the investigation. That strategy tried to highlight the interdependence of water, to insist that the Lake Company's water management improved the overall manufacturing potential of the Merrimack. The ecological interdependence of the river – the fact that the same water regulated by the company inevitably reached factories all along the Merrimack – could be used, Storrow realized, to suit the company's ideological ends.

[57] "The Lake Company Investigation," *LVT*, 23 Feb. 1878.
[58] "Hearing," 2 Mar. 1878.

To strengthen the impact of this point, Storrow proposed in 1877 to admit the Amoskeag Company in Manchester to a third share in the Lake Company venture. The Amoskeag Company had rejected such an offer in 1845. Now the stakes were higher, at least for the corporations at Lowell and Lawrence. As Charles Storrow explained the idea to T. J. Coolidge, the Amoskeag Company's treasurer, the decision to split the Lake Company's stock into three equal shares would benefit the mills at all three cities.[59] For the waterpower companies at Lowell and Lawrence, however, the agreement could have immense value as a public relations ploy. Admitting the New Hampshire company, in Storrow's opinion, "would remove a great deal of the prejudice against them [the PLC and the Essex Company] as strangers, which adds much difficulty & expense of management, places the Lake Co. in a disadvantageous position before local courts & juries, & keeps up a constant call for Legislative influence, which is always dangerous and may be fatal to the present system."[60] By December 1878, the three companies agreed to share equally in the Lake Company.[61] Restructuring ownership of the company sent a potent message: To strike out at the Lake Company would jeopardize the huge manufacturing interests not only in Massachusetts, but in New Hampshire as well.

When Storrow began composing his address before the commission in March, the above agreement had yet to be worked out. But this did not stop him from stressing the value of what the Lake Company had done for manufacturing in New Hampshire. As he put it, since

> all the improvements made by the Lake Company upon all the tributaries of the Merrimack are above the town of Franklin [New Hampshire], they are available at every fall upon the 60 miles of that River between Franklin and the Southern boundary of the State. Upon those sixty miles are situated the largest & most important manufacturing establishments of New Hampshire, to whom these improvements are of at least as great advantage as they are or can be to any beyond the limits of the State.[62]

---

59 C. Storrow to T. J. Coolidge, 15 Oct. 1877, vol. A-17, file 82, PLC Papers.
60 C. Storrow, Memo, 16 Feb. 1878, item 204, EC Papers.
61 George Waldo Browne, *The Amoskeag Manufacturing Co. of Manchester, New Hampshire: A History* (Manchester, 1915), 128. The AMC later reneged on the agreement. See T. J. Coolidge to C. Storrow, 20 Oct., 1 Nov. 1879, item 204, EC Papers.
62 C. Storrow, Remarks on Specifications of Petitioners for Repeal of Lake Co. Charter, 8 Mar. 1878, item 204, EC Papers.

The Lake Company, Storrow continued, "do not [*sic*] draw water *out* of the State, but draw it to run *through* the State; and no earthly power can prevent Merrimack River and all that flows therein from crossing the State boundary line."[63] This was a powerful argument employed by Storrow to absolve the company of any wrongdoing. The manufacturing interests in New Hampshire and Massachusetts were indeed knit together by the interdependence of the river system. By threatening the Lake Company, Storrow cleverly implied, the entire network of manufacturing along the river might unravel.

If repealing the Lake Company's charter imperiled New Hampshire's economic base, allowing the company to continue clearly risked other interests. Storrow could argue persuasively that manufacturing in both states depended on the Lake Company's fastidious control of water. But what of those who resorted to water for other reasons? As the petitioners claimed, the company not only hurt the business economy of the village, but also threatened navigation on Lake Winnipesaukee, farming along the Merrimack, and the public health of the region's citizens.

Of all the petitioners' charges, the public health issue attracted the least attention. It surfaced intermittently at the hearings, but only the local physician, Oliver Gross, appears to have testified on the topic. It was hard to demonstrate any direct connection between the Lake Company's management of the water and the spread of disease. The petitioners never fully fleshed out their argument, but their concerns probably were voiced as follows. Streamflow generally provided adequate year-round dilution of pollution and waste. But when the company held back the flow of water, debris along the banks and streambed probably became exposed. The sight and smell of decomposing waste may have led to alarm over the threat of disease and infection.[64] Storrow, however, quickly dismissed the public health issue. He glibly argued that in the summer and fall, "when the alleged injury to health, if it existed, would

---

63 Ibid.
64 I am inferring here from what happened along the Charles River. During the 1890s, the streambed of the Charles had become strewn with sewage. When the water level was reduced during dry summers, the waste became exposed, producing a bad odor. Physicians later became worried that the low-water conditions and the exposed waste created a favorable environment for malaria to spread: Mills holding back water for power along the river may have made the problem worse, adding to the fears of disease.

be most prominently apparent, very large numbers of people from other states seek these shores at Centre Harbor, at Wolfeboro, at Alton Bay [all towns along Lake Winnipesaukee], and many other places, expressly for the purposes of health and recreation."[65]

Far more attention was devoted – both at the hearing and by Storrow – to whether the company impaired navigation on Lake Winnipesaukee. Steamboat pilots and engineers testified that the company interfered with navigation by drawing down the water in the lake. To counter such views, Storrow designed his response in such a way as to make nature itself the culprit. In his view, "by far the largest part of the fall in the Lake in summer seasons or other times of drought is the result of natural causes, and not of the operations of the Company." The company, Storrow insisted, could not be blamed "for the evaporation from the Lake, or the drying up of the smaller streams or other waters which feed it." Storrow introduced a series of tables to show that the amount of water drawn by the Lake Company could not completely account for the extreme variation in the depth of the lake. And in any case, Storrow urged the commission to "remember how many wheels this very water keeps in motion on its way down to the Southern boundary of the State, & think for a moment what the complaint would have been if it had *not* been so drawn."[66]

Finally, there remained the issue of whether the company had damaged meadows along the Winnipesaukee and Merrimack rivers. The Lake Company interfered at times with farmers harvesting hay by sending water into meadows. Storrow freely admitted that the company changed the "natural condition" of the waters. The point of constructing a dam, he explained, was to do just that, to make a river's flow more uniform so it could be better used for manufacturing. Unregulated water could lead to floods that wasted waterpower, or to paralyzing droughts that might "check or wholly stop the labor of thousands who depend upon it for their daily bread." By constructing dams, he added, mill owners were counteracting the adverse ecological effects of deforestation. Cutting down the forest had made the region's rivers more prone to variation. As Storrow explained:

> The only way in which this natural result of the progress of civilization can be counteracted, is by an intelligent use of other means

[65] Storrow, Remarks on Specifications of Petitioners.
[66] Ibid.

which the progress of civilization also furnishes. Thus when the natural reservoirs fail, artificial reservoirs must take their place & while on the one hand the farmers and lumber men are removing the forests, the mill owners in every direction are husbanding the resources of their ponds & streams.[67]

Storrow portrayed the Lake Company as public minded in its attempts to make the flow more uniform. Here again, its interests in managing the water were "identical with those of the whole manufacturing industry of the State of New Hampshire."[68]

Dismantling the company's system of water control, Storrow concluded, "would be a disastrous blow to the manufacturing industry of New Hampshire." He believed the driving force behind this latest attack on the company came mainly from the town of Lake Village, a place embittered by its failure to become a large industrial center. The waterpower of the town, Storrow contended, would never have supported large-scale industry. Moreover, the persistent attacks on the company only hurt the town's chances of attracting future industry.[69]

Perhaps some people, as Storrow suggested, resented the town's failure to develop into a major industrial city. Others, however, were vexed by the company's grip on the region's water resources. The Lake Company's far-reaching system of water control, however much it improved the overall waterpower of New Hampshire, had concentrated much of the state's substantial water wealth in the hands of a single company. For thirty years, that company had expertly managed the water to aid the mills at Lowell and Lawrence. To be sure, the intrusion of out-of-state capital and a monopoly over water provoked outrage at the company. Yet there still remained the question of who the commissioners would side with in this battle.

The proceedings at Lake Village adjourned on 7 March 1878. They were to begin again at Wolfeborough (Wolfeboro), New Hampshire, two months later to consider the company's water management at Lake Wentworth.[70] In the interim, the chairman of the commission, Francis Faulkner, fell ill and could not

[67] Ibid. The effect of deforestation on streamflow is discussed on p. 106.
[68] Ibid.
[69] Ibid.
[70] Although the petition specifically mentioned both Lake Village and Wolfeborough as places where the Lake Co. had hindered the local economy, the commission seems never to have met in the latter place.

attend the hearings when they began in May. The hearings were thus postponed and the entire matter pushed over into the following year. When Jotham Hutchinson informed James Francis of this news, he also assessed the company's prospects. In Hutchinson's view, the company could rest easy: "I don't think the Lake Co. will have much to fear as to the final result of it [the hearings], unless it is lawyer's fees."[71]

He was right. In the spring of 1879, the commissioners made their final report.[72] Besides the testimony taken at Lake Village, the commission had heard witnesses from the New Hampshire towns of Franklin, Laconia, and Tilton (on the Winnipesaukee River), Ashland (at the Squam lakes), and Bristol (at Newfound Lake). That testimony, it appears, spoke favorably of the Lake Company. Witnesses along the Winnipesaukee, most of whom were factory owners (including George W. Nesmith, former jurist and manufacturer at Franklin), felt satisfied with the company's handling of the water. At Ashland and Bristol, where no one came forth to sign the petition against the company, manufacturers appeared equally content.[73] Although the company had a long and embittered history in the region, much of that conflict had apparently been resolved by 1879. Whatever resentment remained, according to the report, centered in the town of Lake Village. On this score at least, the commissioners agreed with Storrow.

The report itself lacks the air of balance and objectivity we expect of government documents written in our own century. It instead has a flavor decidedly promanufacturing and reads more like a piece of promotional literature urging further economic development of the state's waters. The commissioners found that the Lake Company controlled over one hundred square miles of water. That water, they believed, could "contribute immensely to the wealth and importance of the inhabitants of our own State before it passes into Massachusetts."[74] Moreover, the commissioners were content with the Lake Company's handling of this valuable resource. At the time, the Lake Company had no pending lawsuits, and the verdicts of past

---

[71] J. P. Hutchinson to J. Francis, 28 May 1878, A-48, file 296, PLC Papers.

[72] Faulkner died on 22 May 1879, and John Farr took over as chairman. Henry P. Rolfe, a Concord, New Hampshire, lawyer replaced Faulkner.

[73] *Report of Commission to Consider Matters Relating to the Winnipesaukee Lake Cotton and Woolen Manufacturing Company*, 20 June 1879, pp. 10, 15, a copy of which is in vol. A-22, file 118, PLC Papers.

[74] Ibid., 11.

cases showed that it had paid quite liberally for any damages it had caused.[75] But the central issue was whether the company had stifled the manufacturing potential of New Hampshire. To determine this, the commission compiled figures for six towns fed by Winnipesaukee, Newfound, and Squam lakes (Lake Village, Laconia, Tilton, Franklin, Ashland, and Bristol). With 57 factories, capitalized at over $1.3 million, employing 2,299 people, making use of over 4,700 horsepower of water, and annually producing $5.1 million worth of products, the figures suggested impressive economic progress. The commission saw fit to add in the industry at Manchester and Hooksett, New Hampshire, since it too "largely benefited by the holding of the waters in these reservoirs for the dry seasons."[76]

In the face of such overwhelming achievement, the case against the company failed to impress the commissioners. They dismissed the attack on the company, made mostly in Lake Village, in a single paragraph. Only the complaints of navigators concerned them, and here the commission determined that the company had made up for the damage it had done. While the company had at times lowered the level of Lake Winnipesaukee, it had also paid for the extension of wharfs to accommodate steamboats. And by virtue of the company's excavations at the head of Paugus Bay (Long Bay), small steamboats could now pass from the lake into the bay.[77] Had the controversy been "between the citizens and manufacturers of this State and Massachusetts, and one or the other must suffer in the management of our waters, we should say our duty, as well as our interests, would call upon us and the law-makers of this State to look out for our own welfare first."[78] But at least for the moment, the opposition seemed localized and the overall interests of economic growth well served by the Lake Company. The company's management of the water, the report concluded, "inures to the advantage of our material interests, at the same time Lowell and Lawrence are benefited, and our interests are mutual and run in the same channel."[79]

The commission, it hardly need be said, saw no reason to repeal the company's charter. If anything, just the reverse was

[75] Ibid., 12.
[76] Ibid., 13–14.
[77] Ibid., 17.
[78] Ibid., 16–17.
[79] Ibid., 17.

needed. In a remarkable move, the commission urged the legislature to pass a law amending the mill act of 1868. That law, it will be recalled, gave mill owners the authority to flow lands on *non-navigable* streams provided damages were paid to affected landowners.[80] According to the commissioners, the exception of navigable waters from the law applied to the waters of Lake Winnipesaukee.[81] They recommended legislation allowing the Lake Company the same rights to flood land that the act of 1868 granted other corporations and persons in the state, provided the interests of navigation were safeguarded. Such a law would give the Lake Company more freedom to control water without the fear of litigation from landholders. The commissioners dispelled the notion that their recommendation came at the "request or intimation of the Lake Company." It had occurred to them independently, "on account of the great advantages that would inevitably result to the manufacturing interests on the Winnipesaukee and Merrimack rivers, in this State."[82]

An outcome more favorable to the Lake Company is hard to imagine. It was a long road to this point, full of bitter, hard-fought battles. But in the end, the state commission landed firmly on the company's side.

Over the years, the Lake Company had contrived to alter the waterscape to bring it into line with the needs of the Massachusetts factories. In pursuit of this goal, the company had, along the way, pushed and shoved enough to antagonize many people who lived nearby these great waters. At times that resentment boiled over – sometimes out of reprisal, sometimes out of mere necessity – as those who found their interests compromised took legal action against the company. There was even a brief time when the company, harassed and running scared, had second thoughts about staying and considered turning over the empire of water it had struggled so hard to build. But by the end of the 1870s, much of the commotion had died down, perhaps because the company had thrown enough water in the way of New Hampshire industry. It is not necessary to read closely the 1879 report to sense the commissioners' enthusiasm for production. The commissioners left no ques-

[80] See the discussion in Chapter 5.
[81] Actually, the exception would have applied to the Winnipesaukee River since lakes were not covered under the act.
[82] *Report of Commission*, 17–18.

tion about their infatuation with water as a source of economic potential. What they mainly believed – as did the Lake Company and others who depended on water for energy – was precisely what had fueled the vast transformation of the waterscape. They favored the control of water to serve an industrial calculus; indeed, this was the basis of their thoughts. To them, the Lake Company had committed no sins.

There were others, however, who perhaps felt otherwise. At least this is a conclusion one might draw from the large number of people who petitioned to repeal the Lake Company's charter. The report left unanswered the question of why, if the case against the company had been simply of concern to Lake Village, people across two New Hampshire counties signed the complaint. Since the petitions have not survived, we may never learn the answer to this question. But we can wonder why the petitioners' charges were so readily shuffled aside by the commission. Why did the petitioners lose?

It is seldom safe to speculate on the inevitability of an event. History is rarely so simple. But for a moment, consider the nature of the petitioners' charges. Although the petition held the company responsible for a range of problems, their main complaint was economic in substance. They accused the company of not using the water it controlled for producing textiles in New Hampshire. In the words of the petition, the company had managed its affairs so "as to oppress and crush out the manufacturing interests of various localities on the borders of Lake Winnipesaukee."[83] There was no challenge to the company's right to use water for production. Instead, the petitioners questioned who would reap the benefits of using water to produce. Beneath their principal complaint lay an unquestioned assumption: Water was meant to be used for instrumental purposes, to maximize economic gain. They framed the debate in terms of whether or not the company contributed to the economic development of the region. That was a framework in which water figured primarily in terms of its value to production.

Having cast the debate in such a fashion, the petitioners left the Lake Company a chance at vindication. Few knew the productive value of water better than it did. All it need do was show that it improved the overall waterpower of New Hampshire –

---

[83] The final report contains in it a copy of the actual charges made by the petitioners. See ibid., 4.

and it did that well. Here it was helped by the interdependent nature of the water itself. The Lake Company claimed, and not unfairly, that its efforts benefited not merely its own interests, but those of manufacturers throughout New Hampshire. That argument had enormous appeal to a society transfixed by production. And it alone may well have been potent enough to exonerate the company.

EPILOGUE

When Henry Thoreau died in 1862, the Merrimack River was still on his mind. On the day of his death, his sister, Sophia, read aloud to him from *A Week on the Concord and Merrimack Rivers*. She read the sentence "We glided past the mouth of the Nashua, and not long after, of Salmon Brook, without more pause than the wind." Thoreau then replied, "Now comes good sailing."[1] Moments later, he was dead, whisked away from a world in love with progress – a world he devoted a lifetime to criticizing.

Thoreau died at the age of forty-four. Had he lived on, would he have commemorated the fiftieth anniversary of his trip along the Concord and Merrimack rivers? We might imagine the old curmudgeon, were he sturdy enough to travel, venturing off again into the waters of New England. On the one hand, some of the developments taking place in 1889 probably would have heartened Thoreau. That was the year the Lake Company was formally dismantled. The PLC and the Essex Company sold their shares in the company to a group of New Hampshire manufacturers.[2] The same decade, both Charles Storrow and James Francis retired from their respective positions at the Lowell and Lawrence waterpower companies. Each of them had devoted almost half a century to the mills along the lower Merrimack River, to experimenting, regulating, distributing, and creating waterpower. Now their work was at an end. By the 1880s, the Waltham–Lowell system was all but finished in New England – a development Thoreau would have relished.

---

[1] This anecdote is recounted in Linck C. Johnson, "Historical Introduction," in Henry D. Thoreau, *The Illustrated "A Week on the Concord and Merrimack Rivers"* (Princeton, N.J., 1983), xxxvi, 351.

[2] "Improving the Water Supply," *Boston Morning Journal*, 27 Sept. 1889. The new owners were expected to control the water to benefit manufacturing along the Winnipesaukee River.

On the other hand, there were signs in the 1880s of the vitality of the industrial mode of production. During the early part of the decade, plans were made for a new textile city at Sewall's Falls in Concord, New Hampshire. At that point, the Merrimack River dips nineteen feet before leveling out and winding through a low-lying tract of land. The Concord Land and Water Power Company, incorporated in 1881, purchased property and water rights at the falls. Significantly, the founders of the company were not Boston Associates, but New Hampshire manufacturers.[3] Most of the original promoters of the Waltham–Lowell mills, figures such as Francis Cabot Lowell, Patrick Tracy Jackson, Abbott Lawrence, and Nathan Appleton, had been dead for some time.

The organizers of the new Concord company did, however, call on the able Waltham–Lowell engineer, Hiram Mills, to help them develop the waterpower. Between 1881 and 1884, Mills designed a granite dam and a power canal spanning 1,800 feet in length. The technical apparatus would supply 22 feet of fall at the production sites, providing fifty-five mill powers. According to the company's promotional literature, the future of waterpowered manufacturing rested in Concord, "the next place in natural order . . . [where] manufacturing can be done more economically by water-power than by steam-power."[4]

The language used here is very suggestive. In the company's view, the natural features of Concord made it a logical choice for waterpowered manufacturing. As it wrote: "Nature has long since done her part – now it is man's opportunity. The present is unquestionably the time to make one more draft on the resources of the Merrimack."[5] In fact, the plans to make Concord a major center of waterpowered manufacturing never materialized. When the dam and power canal were finally finished in 1894, the company furnished electricity, not waterpower, to the city.[6]

[3] George B. Lauder, "The Story of Sewall's Falls" (ms., 1938), Concord Room, Concord Public Library, Concord, N.H., 4.

[4] *Proposed Development at Sewall's Falls on the Merrimack River, in Concord, N.H., by the Concord Land and Water Power Company, September 15, 1884* (Concord, N.H., 1884), 5, 9–10, 18.

[5] Ibid., 24.

[6] See "First Annual Report of the Directors of Concord Land and Water-Power Co. to its Stockholders, January 1, 1894," N.H. State Library, Concord, N.H. This was the same dam that devastated efforts to restock the Merrimack with salmon, discussed on p. 201.

Still, whether falling water generated waterpower or electric, both uses hinged on the control of nature. Over the course of the nineteenth century, the mastery of the natural world became inextricably tied to a desire for progress. As the century came to a close, New Englanders demonstrated a growing faith in their ability to manipulate and exploit the environment. The control of nature was beginning to harden into custom and convention. To be sure, greater command over water, over nature in general, had its positive points – rising living standards for some, more comfort and convenience. But there is a troubling side to this aggressive, manipulative posture toward the natural world – a problem that penetrates to the core of modern American culture. Inspired by the technical triumphs of the twentieth century, we have been lulled into thinking that nature can be dominated at will, seduced by our seemingly invincible ability to conquer the environment. It is worth pausing to consider the truth of this proposition. This unquestioned attitude of dominance toward nature is among the most powerful legacies of industrialization. It is also an attitude we may rightly suspect to be little more than an illusion.

A NOTE ON SOURCES

This study rests on an eclectic set of sources. To a large extent, the sources are the typical ones employed by historians. I have tried, however, to read these traditional sources for their environmental import. Here I will briefly indicate those primary sources I found most valuable in writing about water and industrial transformation. For a comprehensive bibliography, see my unpublished doctoral dissertation, "Nature Incorporated: The Waltham–Lowell Mills and the Waters of New England" (Brandeis University, 1989).

BUSINESS RECORDS

Information on waterpower at Lowell can be found in the records of the Proprietors of Locks and Canals on Merrimack River, at the Baker Library, Harvard Business School. Likewise, the Essex Company papers at the Museum of American Textile History (MATH) provide substantial information on all aspects of waterpowered textile production at Lawrence. I consulted these collections to determine the early business corporation's distinct relationship with nature. The papers of James Francis and Charles Storrow were especially useful for understanding the technology of water control. Directors' and treasurers' reports substantiated when certain water control devices, such as dams and power canals, were put into operation.

No separate set of records exists for the Lake Company. But extensive correspondence between the Lake Company's agents and the engineers at Lowell and Lawrence is available in the above collections. I was able to reconstruct the conflicts over water from this correspondence. In addition, Charles Storrow's diary, in the Essex Company collection, helped flesh out the Lake Company's strategy for controlling the New Hampshire lakes.

The records of the Amoskeag Manufacturing Company are divided between two archives. Directors' and treasurers' records are available at Baker Library. Additional Amoskeag papers, including engineering records, can be found at the Manchester Historic Association. At Baker, the papers of the Boston Manufacturing Company, Merrimack Manufacturing Company, and Nashua Manufacturing Company were also helpful.

PERSONAL PAPERS

The Baldwin Collection at Baker Library includes the correspondence and papers of Loammi Baldwin, Loammi Baldwin, Jr., and James Baldwin. The members of this remarkable family of engineers were involved in many water control projects, and the collection reflects the breadth of their interests. I also made use of a small collection of the papers of Loammi Baldwin, Jr., at the University of Chicago Library.

Personal papers for some of the Boston Associates are available at the Massachusetts Historical Society. Although the papers rarely deal directly with matters regarding the natural world, they are at times enlightening. I made use of the Appleton Family Papers and the Francis Cabot Lowell Papers.

LEGAL DOCUMENTS

Nineteenth-century water law is discussed in Joseph K. Angell, *A Treatise on the Law of Watercourses,* 5th ed. (Boston, 1854), and idem, *A Treatise on the Right of Property in Tide Waters and in the Soil and Shores Thereof,* 2d ed. (Boston, 1847). Nathan Dane, *A General Abridgment and Digest of American Law,* 9 vols. (Boston, 1823–9), and James Sullivan, *The History of Land Titles in Massachusetts* (Boston, 1801), were also of value.

The published New Hampshire and Massachusetts water law cases were essential to this study. I tried to read these cases both for their legal substance and for what they indicated about how jurists and others perceived the natural world.

Aside from the published case law, I also relied on the unpublished records of New Hampshire's highest court (variously called the supreme court of judicature and the supreme judicial court). The records, available at the New Hampshire Supreme Court Library, are indexed and contain a range of documents including handwritten briefs, memos, and letters. I used the records to explore the social and ecological relations behind the more abstract legal rulings. The collection is invaluable for understanding how class interests influenced the formation of legal doctrine. Those interested in unpublished court records in Massachusetts should see Michael S. Hindus, "A Guide to the Court Records of Early Massachusetts," in *Law in Colonial Massachusetts 1630–1800: A Conference Held 6 and 7 November 1981 by the Colonial Society of Massachusetts* (Boston, 1984), 519–40.

Papers regarding acts and resolves passed in colonial Massachusetts are available at the Massachusetts Archives at Columbia Point. I relied on petitions filed in support of certain legislation to unpack the history of conflict over water on the Charles River.

To understand how water rights were exchanged, I turned to county registries of deeds in New Hampshire and Massachusetts. Working with land deeds is laborious, but often extremely rewarding. By tracing property transactions, I was able to examine the Lake Company's attempt to secure control of New Hampshire's lakes. In addition, consulting the PLC's deeds helped me to understand how waterpower was leased.

GOVERNMENT DOCUMENTS

The published reports of the Tenth Census of the United States (1880) are a remarkably rich historical resource. Volumes 16 and 17, entitled *Reports on the Water-Power of the United States* (Washington, D.C., 1885–7), survey waterpower sites throughout the country. George F. Swain's report, "Water-Power of Eastern New England," offers a detailed examination of water control technology in the Merrimack valley.

The published reports of several state agencies were used to determine the environmental consequences of industrial change. In particular, the reports of the Massachusetts Commission on Fisheries (later called the Commission on Inland Fisheries) provided important information on fish restoration. The reports of the Massachusetts State Board of Health discuss pollution, water supply, and waterborne disease. I made somewhat less use of the reports of the New Hampshire Commission on Inland Fisheries and the State Board of Health for the State of New Hampshire, although there is much valuable material to be found in them.

The published *City Documents of the City of Lowell* were helpful for exploring the development of Lowell's public water supply during the 1870s. On water supply and sewers in Lawrence, see L. Frederick Rice, "Engineer's Report," in *Report on Proposed Water Works at Lawrence, Mass.* (Lawrence, 1872), and *Report on a General System of Sewerage for the City of Lawrence* (Lawrence, 1876). For how the city of Manchester handled these issues, see "Report on Sources of Water Supply, for the City of Manchester, N.H., with Estimates of Cost," in Board of Water Commissioners of the City of Manchester, New Hampshire, *First Annual Report* (Manchester, 1873), and J. T. Fanning, "Engineer's Report," in Board of Water Commissioners of the City of Manchester, New Hampshire, *Third Annual Report* (Manchester, 1874).

OTHER SOURCES

Local histories offered detailed information on industrial transformation in the Charles and Merrimack valleys. See George W. Chase, *The History of Haverhill, Massachusetts* (Haverhill, 1861); Charles Cowley, *Illustrated History of Lowell*, rev. ed. (Lowell, 1868); Samuel Adams Drake, ed., *History of Middlesex County, Massachusetts*, 2 vols. (Boston, 1880); D. Hamilton Hurd, *History of Merrimack and Belknap Counties, New Hampshire* (Philadelphia, 1885); Daniel Lancaster, *The History of Gilmanton* (Gilmanton, N.H., 1845); James O. Lyford, ed., *History of Concord, New Hampshire*, 2 vols. (Concord, N.H., 1903); C. E. Potter, *The History of Manchester, Formerly Derryfield, in New Hampshire* (Manchester, 1856); M. T. Runnels, *History of Sanbornton, New Hampshire*, 2 vols. (Boston, 1882); and H. A. Wadsworth, *History of Lawrence, Massachusetts* (Lawrence, 1880).

Descriptions of New England's land and waterscape are available in gazetteers and travelers' accounts such as John Warner Barber, *Historical Collections, Being a General Collection of Interesting Facts, Traditions, Biographical Sketches, Anecdotes, & c. relating to the History of Antiquities of Every Town in Massachusetts, with Geographical Descriptions* (Worcester, Mass., 1839); Timothy Dwight, *Travels in New England and New York*, ed. Barbara Miller Solomon, 4 vols. (Cambridge, Mass., 1969); Nathan Hale, *Notes Made During an Excursion to the Highlands of New-Hampshire and Lake Winnipiseogee* (Andover, Mass., 1833); and John Hayward, *The New England Gazetteer Containing Descriptions of All the States, Counties and Towns in New England*, 2d ed. (Concord, N.H., 1839).

INDEX